Mathematical Analysis: A Translation

Alex Chaney

August 6, 2017

Preface

> Theory is crucial. Serendipity may occasionally yield insight, but it is unlikely to be a frequent visitor. Without theory, we make endless forays into unchartered badlands. With theory, we can separate fundamental characteristics from fascinating idiosyncrasies and incidental features. Theory supplies landmarks and guideposts, and we begin to know what to observe and where to act.
> –John H. Holland in Hidden Order, How Adaptation Builds Complexity

Reading math is usually hard work, and every mathematics author struggles with how much to write. If you write too much, some smart folks will wonder why you're wasting their time with too much detail. If you write too little, the audience (usually students) are lost or struggling. My goal is to supplement a textbook whose author is probably a researcher in Analysis and who has tended to write less explanation than what a novice student is yearning for. A related issue is what do I expect the audience to know?

I expect the audience to know what is often covered in a math course usually with "Foundations" or "Proof" in the title. You have done a few induction proofs, subset proofs, and a few by contradiction, but you need to see some more examples. You know what has to be done to prove two statements are equivalent. You have seen some truth tables, but probably don't want to revisit them. In a crunch you can look up how to form the contrapositive of a statement with a couple quantifiers like 'there exists' or 'for all' and be comfortable moving on. You have taken a course or two in Calculus, enough to become comfortable with algebra and inequalities. Your calculus experience included the guess and check method for solving a differential equation. You have probably not taken another course that requires proof writing such as Abstract Algebra, Number Theory, Topology, or Abstract Linear Algebra.

We are going to present the distilled version of what took 2205 years to develop (from the birth of Archimedes in about 287 BC to the death of Georg Cantor in 1918). I take no credit for any of that distillation process...my contribution is the arrangement and presentation that is hopefully useful to a modern student.

The presentation starts out in a very pedantic style, taking much more time with preliminaries that usually get little to no coverage in the classroom, yet remaining concise and with a mathematical style of writing. This is done purposely to develop a deliberate thought process. Much of writing a proof is deciding what is to be taken for granted, and the novice should take very little for granted. The beginning includes much more commentary (after the preliminaries) than the latter parts of the book, but it is the combination of the commentary with the proof that hopefully achieves success in communicating the process of discovery–how to think about math analysis. In the later chapters, we anticipate the growth of the reader by including less discussion about the proof technique, and more outline of the links between the theorems that are presented. Also, the last chapter has much more commentary to relate math analysis to related topics or courses.

I have been careful to present the material in a logical order that does not refer to any result not established previously. Of course there are a couple transgressions, like assuming a non empty set of natural numbers contains its minimum, and a couple properties from topology that must be taken for granted without an introductory course, or a long diversion.

I have omitted many of the accoutrements of a textbook, most glaringly is the lack of exercises. I have ventured to do the opposite, to present more solved examples, and commentary. I think most undergraduate students need a maximum of examples and explanation in addition to some problems that they have to struggle with and solve on their own. In the internet age it is too easy for a student to look up answers for textbook problems with the authors name and problem number. Indeed, an ambitious teacher could supplement this book with some old exercises from the vast collection of analysis books already on the market, and challenge even the most resourceful students.

The material covered mixes examples that are mostly exercises or examples from other textbooks with the definitions and theorems necessary to get to and through the fundamental theorem of calculus quickly, yet thoroughly. As a result, the more interesting topics that branch out to applications, and other courses come at the end. The selection of topics coming after the fundamental theorem are chosen to develop the student to understand the Dominated Convergence Theorem (but not its proof). This is done through a series of examples that answer questions about convergence, and illuminate some of the subtle behavior of sequences of functions. The capstone is a rigorous application of power series to ordinary differential equations and the interesting behavior that branches to complex analysis, or to test functions for distributions.

I am especially grateful to Professor Doug McInvale of the United States Military Academy for selecting me to teach the math analysis course there. Also, Jocelyn Bell and Joe Pedersen were great support when my understanding of analysis developed. The greatest contribution came from Amy Givler, and her comments on an early version of the manuscript.

Finally, while much effort has been exhausted to eliminate errors, I am sure there are many. Since the text is printed on demand I am able to make corrections and implement them within a couple work days. An Errata page will be posted on Facebook, where corrections can be submitted for review.

Contents

Chapter 1

The Axiomatic Approach

1.1 Natural Numbers

Our first use of numbers is for counting, and that is where we will start, but in a more rigorous way. By rigor, we mean that we start with a few given statements, and make strictly logical conclusions from them. However, we should choose our axioms so that our intuition is useful. The intuition is explained on the left side, while the opposite page contains a concise way of saying the same thing in mathematical notation.

N1 We can think of counting as starting with one, but then our system would have no way of representing the absence of an object. Therefore we use the symbol called zero to represent the absence of something.

N2 As we count, each number comes after the previous. So for any number, say n, it has a succeeding number, $n + 1$. We write this as $s(n) = n + 1$, and call $s(n)$ the successor of n, or in specific cases we give it the usual name for the number. For example $s(97) = 98$.

N3 Since we want to include the notion of the absence of an object, represented by 0, we start counting with zero. This is a somewhat arbitrary choice, but we have to start somewhere, so it is with 0. Therefore, 0 cannot the the successor of any other element. Otherwise we would be starting somewhere else.

N4 If I have four quarters in my pocket, and you have four quarters in your pocket, then we need a way to say that we have the same number of quarters. We do this as follows. If the successor of the number of quarters in your pocket is $s(n) = 5$ and the successor of the number of quarters in my pocket is $s(m) = 5$, then we have the same number of quarters $n = m = 4$.

N5 The Principle of Mathematical Induction is a tool to say something is true about all the natural numbers. A famous example is that there are an infinite number of integers that are prime numbers. Our first example will be that the sum of the first n integers is $\frac{n(n+1)}{2}$. But before we get to the examples, lets take a closer look at what the principle says on the next page.

This last item stands apart, but in a way similar to the axioms of plane geometry from Euclid. The first four axioms of Euclid can be stated in about ten English words each, and the fifth takes about forty words. This is probably just a coincidence with the last axiom of the natural numbers.

The Peano Axioms for Natural Numbers, $\mathbb{N}$.

N1 $0 \in \mathbb{N}$.

N2 For every $n \in \mathbb{N}$, $s(n) = n + 1 \in \mathbb{N}$.

N3 For any $n \in \mathbb{N}$, $s(n) \neq 0$.

N4 $s(n) = s(m) \to n = m$.

N5 For $X \subset \mathbb{N}$, if a first element, $0 \in X$ and $n \in X \implies s(n) \in X$, then $X = \mathbb{N}$.

If we have just two criteria met for some property, P_n, that we can relate to the integers, we'll conclude that the property holds for all the integers, and say P_n is true for all $n \in \mathbb{N}$.

We need a first element that is true. With zero as our starting point, this will often not be the case. Consider the statement $n^2 > 2n$. It is not true for $n = 0, 1, 2$, but it turns out to hold for $n \geq 3$. On the other hand, the sum of the first n integers turns out to be $\frac{n(n+1)}{2}$ and this holds for $n \geq 0$. This first criteria is often called the base case.

The second criteria that we need is that a statement about n implies that the same statement about $n + 1$ is true. Note that we should not interpret this to say that if the first domino reaches the second, then the rest of the dominoes will fall. It should be interpreted to say that if an arbitrary domino reaches the next one in line, then all the subsequent dominoes will fall. This second criteria is often called the induction assumption.

These two criteria are put together, usually with some creative algebra, in order to form a proof by induction. The necessary steps are as follows:

1. Verify the base case is true for the selected starting point, usually zero or one.

2. Use the statement P_n to show $P_n \to P_{n+1}$.

3. By the principle of mathematical induction we conclude the statement is true for all n greater than or equal to the base case.

1.1.1   Induction

Example 1   Prove $n^2 > 2n$ for all true cases.

Discussion: We start with the inequality in question

$$n^2 > 2n$$

For $n = 0, 0 > 0$ is false. For $n = 1, 1 > 2$ is false. For $n = 2, 4 > 4$ is false. For $n = 3, 9 > 6$ is true, so we start with this case. Next we write one side of the statement we want to prove and manipulate it until we get the desired result. Along the way, we have to use the induction assumption, the statement P_n which in this problem is $n^2 > 2n$. We can use other facts as well, but we must use the induction assumption (otherwise it is not a proof by induction). Our goal is the statement P_{n+1} which in this example is $(n + 1)^2 > 2(n + 1)$. Once we have verified the base case and used P_n to show P_{n+1}, we let the reader know we are using the principle of mathematical induction to conclude the result.

Example 2   Prove that the sum of the first n integers is $\frac{n(n+1)}{2}$.

Discussion: For this example the base case is true for $n = 0$ as shown in the proof. We know what we want the result to be, just replace n with $n+1$ in the right hand side of the induction assumption. That is we want to show

$$0 + 1 + 2 + \cdots + n + (n + 1) = \frac{(n + 1)((n + 1) + 1)}{2}.$$

But we cannot just write this without justifying it! We first use the induction assumption which gives us $\frac{n(n+1)}{2} + n + 1$ on the left hand side. A technique to use is to write the left side as a polynomial, and then since we know the factors we want on the right side, use long division to factor them out. The left side numerator is then $n^2 + 3n + 2$, and since we want one of the factors to be $n + 1$ we do division

$$\frac{n^2 + 3n + 2}{n + 1} = n + 2$$

This may seem a bit overkill for a second order polynomial, but the point is to give a technique that can work for exercises with higher order polynomials.

Example 1 Prove $n^2 > 2n$ for all true cases.

Proof The base case is $n = 3$. Observe that $3^2 = 9 > 2 \cdot 3 = 6$. Consider that
$$(n+1)^2 = n^2 + 2n + 1.$$
By the induction assumption, $n^2 > 2n$,
$$(n+1)^2 = n^2 + 2n + 1 > 2n + 2n + 1.$$
Since the base case is $n = 3$, we observe that $2n + 1 > 2$ and use that fact to write
$$(n+1)^2 = n^2 + 2n + 1 > 2n + 2n + 1 > 2n + 2 = 2(n+1),$$
hence the induction assumption (with some algebra) implies the statement for $n+1$,
$$(n+1)^2 > 2(n+1).$$
Therefore, by the principle of mathematical induction, $n^2 > 2n$ is true for all $n \geq 3$ $\square$

Example 2 Prove that the sum of the first n integers is $\frac{n(n+1)}{2}$.

Proof For the base case, $0 = \frac{0(0-1)}{2} = 0$. Consider the sum of the first $(n+1)$ integers
$$0 + 1 + 2 + \cdots + n + (n+1).$$
By the induction assumption, $0 + 1 + 2 + \cdots + n = \frac{n(n+1)}{2}$, we have
$$0 + 1 + 2 + \cdots + n + (n+1) = \frac{n(n+1)}{2} + n + 1.$$
And then creatively applying some basic algebra yields
$$\frac{n(n+1) + 2n + 2}{2} = \frac{n^2 + 3n + 2}{2} = \frac{(n+1)(n+2)}{2} = \frac{(n+1)((n+1)+1)}{2},$$
hence
$$0 + 1 + 2 + \cdots + n + (n+1) = \frac{(n+1)((n+1)+1)}{2}.$$
Therefore, by the principle of mathematical induction, the sum of the first n positive integers is $\frac{n(n+1)}{2}$ for all $n \in \mathbb{N}$ $\square$

Binomial Theorem   Use induction and

$$\binom{n}{k} = \frac{n!}{k!(n-k)!} \quad \text{with } n! = 1 \cdot 2 \cdot 3 \cdots n; \text{ and } 0! = 1$$

in order to prove the binomial theorem for any numbers a, b,

$$(a+b)^n = \sum_{k=0}^{n} \binom{n}{k} a^{n-k} b^k.$$

Hint   Apply this useful fact,

$$\binom{n}{k} + \binom{n}{k-1} = \binom{n+1}{k} \quad \text{for } k = 1, 2, \ldots, n$$

which can be verified by applying the definitions at the top.

Discussion: The proof on the opposite page is a significant leap from the previous two in terms of the algebra skills required, yet the structure is exactly the same as most other inductive proofs (verify base case, use P_n to show P_{n+1} is true, cite the principle of mathematical induction).

The pattern of the previous two proofs, that we do not write what we want to show until it is fully justified, is continued. This example uses more algebraic knowledge and the rules to manipulate summation notation, and that makes it more challenging that the previous two. The approach given is based on the following observations:

- The choice to re-index the sums is made based on the identity given in the hint.

- In order to combine two summations into one they must start and stop at the same index values

The use of summation (or sigma) notation makes this problem nice and tidy in comparison to writing the terms of the sum with $\cdots$ in the middle. It is essential to know the rules for sigma notation, and they are included in every calculus textbook (often in an appendix).

Proof   For the base case we have

$$(a+b)^0 = 1 = \sum_{k=0}^{0}\binom{0}{k}a^{0-k}b^k.$$

Next we'll start with $(a+b)^{n+1}$ and use a rule of exponents

$$(a+b)^{n+1} = (a+b)^n(a+b).$$

By the induction assumption we have:

$$\left(\sum_{k=0}^{n}\binom{n}{k}a^{n-k}b^k\right)(a+b).$$

Then use the distributive property to get

$$\sum_{k=0}^{n}\binom{n}{k}a^{n-k+1}b^k + \sum_{k=0}^{n}\binom{n}{k}a^{n-k}b^{k+1}.$$

Re-index the sums using $k = i-1$ on the second sum, and rename $k = i$ in the first sum.

$$\sum_{i=0}^{n}\binom{n}{i}a^{n-i+1}b^i + \sum_{i=1}^{n+1}\binom{n}{i-1}a^{n-i+1}b^{i-1+1}$$

Now, modify the first sum so that the indices start and stop at the same number as the second sum.

$$\binom{n}{0}a^{n+1} + \sum_{i=1}^{n}\binom{n}{i}a^{n-i+1}b^i + \left(\sum_{i=1}^{n}\binom{n}{i-1}a^{n-i+1}b^i\right) + \binom{n}{n}b^{n+1}$$

Then we can apply the identity on the previous page to combine the sums, but at the same time we bring the $n+1$ term inside the summation.

$$\binom{n}{0}a^{n+1} + \sum_{i=1}^{n+1}\binom{n+1}{i}a^{n+1-i}b^i$$

Finally, move the first term in the summation by starting it at $i = 0$.

$$\sum_{i=0}^{n+1}\binom{n+1}{i}a^{n+1-i}b^i$$

Therefore, by the principle of mathematical induction , we conclude that the binomial theorem is true for all integers n $\square$

1.2   Rational Number Field

In the previous section It may have seemed as if we were just using induction, but we actually were doing a lot more. We added, multiplied,
divided and subtracted the natural numbers. From the natural numbers
we can think of the integers (positive and negative) as the natural numbers with their reflection about the point zero. Then if we allow division,
we get numbers like 1/3 which is not an integer, but can be written as the
ratio of two integers. There is an algebraic structure that we use for arithmetic and it is called the field of rational numbers. Just like there are axioms for the integers, there are axioms for the field of rational numbers.
We'll state the field axioms, and prove all the usual algebra operations
using these axioms. There are other fields besides the rational numbers,
but the most widely used one is probably the rational numbers.

The other things we used in the previous examples are equality and
inequality. Formally, the equals sign is a binary relation that is reflexive,
symmetric, and transitive. We omit developing its axiomatic background
and rely on your intuition that two values are equal if they are the same
number. However, we apply an ordering on the rational numbers which
is symbolized by the inequality signs and do this in a formal way.

We take it as obvious that $a - 0 = a$ but this simple statement is an application of two different field axioms. If we start with the left side, $a - 0$,
it actually means $a + (-0)$, that we are adding a to the additive inverse of
0. We have an axiom that tells us that 0 exists with its expected meaning,
and another axiom that says the additive inverse of zero (or any rational number) exists. Combining these axioms in a logical manner yields
the result we expect. However, we need a third axiom, commutativity, in
order to conclude that $0 - a = a$. It is tedious to think about arithmetic
this way, but it is also good practice to avoid sloppy reasoning with other
building blocks when we introduce them.

The Field Axioms for Rational Numbers, $\mathbb{Q}$.

Q1 If $a, b \in \mathbb{Q}$ then $(a + b) \in \mathbb{Q}$.

Q2 If $a, b \in \mathbb{Q}$ then $(ab) \in \mathbb{Q}$.

These are called the closure axioms. They tell us that whatever legal operation we perform on two elements from the field, the result will be an element in the field. Notice that if we multiply 2 with the inverse of 3 we get 2/3 which is not an integer. Therefore the integers are not a field.

Q3 If $a, b, c \in \mathbb{Q}$ then $(a + b) + c = a + (b + c)$.

Q4 If $a, b, c \in \mathbb{Q}$ then $(ab)c = a(bc)$.

These are called the associative axioms. They tell us that the order in which we perform one type of operation does not matter. So we can write a product or sum of more than three terms without parentheses like this, 2+3+4=9. In contrast, note that 4^{2^3} is ambiguous because

$$4096 = \left(4^2\right)^3 \neq 4^{\left(2^3\right)} = 65536.$$

Q5 If $a, b \in \mathbb{Q}$ then $a + b = b + a$.

Q6 If $a, b \in \mathbb{Q}$ then $ab = ba$.

These are the commutative axioms. They tell us that, separately, the order of addition and multiplication does not matter. We know that $2 - 5 = -3 \neq 5 - 2 = 3$, but the commutative property is for the operation of addition. If we carefully write the operations we see that they are $2 + (-5)$ and $5 + (-2)$ which are obviously not the same.

Q7 There exists $0 \in \mathbb{Q}$ such that for all $x \in \mathbb{Q}, x + 0 = x$.

Q8 There exists $1 \in \mathbb{Q}$ such that for all $x \in \mathbb{Q}, x \cdot 1 = x$.

These are the identity axioms. They say that zero is in the field and if we add zero to any number we get the original number, and similarly for one and multiplication. They are useful when developing the rules of algebra that we routinely follow.

Q9 For all $a \in \mathbb{Q}$ there exists $-a \in \mathbb{Q}$ such that $a + (-a) = 0$.

This says that every number has an additive inverse. As a result we have an operation we normally call subtraction but is actually addition of the additive inverse of a positive number. See Q5 above.

Q10 For all $a \in \mathbb{Q} \setminus \{0\}$ there exists $a^{-1} \in \mathbb{Q}$ such that $a(a^{-1}) = 1$.

This says that every number except zero has a multiplicative inverse. This is analogous to Q9 above but it leaves $\frac{0}{0}$ undefined as we are accustomed to. This is the property that makes the rational numbers rational, in the sense that rational numbers are a ratio of two numbers. The ratio (or division) is precisely the product of a number and the inverse of another number, so another way of writing $1 \cdot a^{-1}$ is $\frac{1}{a}$.

Q11 If $a, b, c \in \mathbb{Q}$ then $a(b + c) = ab + ac$.

This Is the distributive property. It relates the operation of addition with the operation of multiplication in the familiar way.

1.2.1 Algebraic Properties of $\mathbb{Q}$

Historically, we humans did not write the axioms and then deduce the rules of algebra. So in the temporal sense the following algebra examples are backwards. However, in recent history we have found that exploiting the general rules has been fruitful for the discovery of new knowledge. One example of this is the study and application of statistics. In order to decide what meaning should come from a set of data we presume that the data is random and apply the principles of random variables to decide if the data is consistent with the theory of random variables. If the data is not consistent with the theory we conclude that some non-random effect is present. If this non-random effect is present throughout a large collection of data, and has just recently been recognized by many researchers, we have discovered new knowledge.

An alternative conclusion would be that the theory of random variables is incorrect or does not apply. Given that so much of our knowledge rests upon the theory of random variables, it would be very uncomfortable to abandon that theory at this point. But in order to build our ability to make new theory it is good practice to apply the known theory, even if the results appear obvious or trivial. We must recognize that learning means to learn to stand on the shoulders of the giants that came before, even if we are to become one of those giants. It is for these reasons that it is worthwhile to study the following examples, or better yet, develop your own proofs. This is an opportunity to explore how some algebraic properties can be built from others. We'll assume the elements we start with are all rational numbers in the following examples. Also, it might be useful to write the field axioms on a sheet of scratch paper, or make a copy of that page, instead of flipping back repeatedly.

Finally, we have developed the notion of using subtraction and division in arithmetic. This is a more efficient way of thought, but notice that the field axioms don't use those terms. We also think of -2 as a negative number, not as the additive inverse of 2. Then if we multiply -2 with another number, when being precise we write -2 as (-2), the additive inverse of 2, not negative two.

Cancellation law for addition   Prove if $a + c = b + c$ then $a = b$.

Proof   Since $c \in \mathbb{Q}$, by additive inverse property there is $-c \in \mathbb{Q}$. By closure we can add $-c$ and the result is still a rational number:

$$a + c + (-c) = b + c + (-c).$$

Then by the additive inverse property we have

$$a + 0 = b + 0.$$

Then apply the additive identity property on both sides to get

$$a = b \; \square$$

Example 3   Prove $x + y = 0$ implies $x = -y$.

Proof   By additive inverse there exists $-y$ such that $0 = y + (-y)$. Then we have that

$$x + y = 0 = y + (-y).$$

By the commutative law we have that

$$x + y = 0 = (-y) + y.$$

Since equality is transitive we have that

$$x + y = (-y) + y.$$

Then apply additive inverse to y on both sides to see that

$$(x + y) + (-y) = (-y + y) + (-y).$$

By associativity we can write

$$x + (y + (-y)) = -y + (y + (-y)),$$

and by the additive inverse we have that

$$x + 0 = -y + 0.$$

Then by the additive identity we have that

$$x = -y \; \square$$

Example 4   Prove $-(-x) = x$.

Proof   By the additive inverse if $x \in \mathbb{Q}$ then there exists $-x \in \mathbb{Q}$. By the additive inverse again, since $-x \in \mathbb{Q}$ there is $-(-x) \in \mathbb{Q}$ such that

$$(-x) + (-(-x)) = 0.$$

Add x to both sides and by associativity we won't add parenthesis

$$x + (-x) + (-(-x)) = x + 0.$$

Then apply additive inverse to the first two terms on the left to get

$$0 - (-x) = x + 0.$$

Then apply commutativity to the left side and then additive identity on both sides to get the result,

$$-(-x) = x \quad \square$$

Example 5   Prove $a \cdot 0 = 0$ for all a.

Proof   Let $c \in \mathbb{Q}$. By additive inverse, $0 = c + (-c)$ which we might write as

$$0 = c - c.$$

Since $a \in \mathbb{Q}$ and $0 \in \mathbb{Q}$, by closure their product is in $\mathbb{Q}$.

$$a \cdot 0 = a(c - c)$$

The distributive law gives us

$$a \cdot 0 = a(c - c) = ac - ac.$$

By closure ac is a rational number, so that the additive inverse applied to $ac - ac$ yields

$$a \cdot 0 = a(c - c) = ac - ac = 0.$$

So by transitivity of equality we have

$$a \cdot 0 = 0 \quad \square$$

Example 6   Prove $(-a) \cdot b = -ab$.

Proof   By additive inverse we have that

$$a + (-a) = 0.$$

Then multiply by b to get

$$(a + (-a))b = 0b.$$

By the distributive property we have

$$ab + (-a)b = 0b.$$

By example 5 we have

$$ab + (-a)b = 0.$$

Then by additive inverse there is $-ab$ that we can add to both sides.

$$(-ab) + ab + (-a)b = -ab + 0.$$

By commutativity we have

$$ab + (-ab) + (-a)b = -ab + 0.$$

By additive inverse we have

$$0 + (-a)b = -ab + 0.$$

By commutativity we have

$$(-a)b + 0 = -ab + 0,$$

and finally by additive identity we have that

$$(-a)b = -ab \quad \square$$

While the above example may seem like a lot of work or excessively tedious, this is what modern mathematics is about. It is about reasoning in a strictly logical manner from an accepted collection of axioms and definitions. Notice that the last three proofs started with an equality that is from one of the field axioms. We can discover which one to use by working backward from the desired conclusion.

Example 7   Prove $(-a) \cdot (-b) = ab$.

Proof   Note that by example 5 we have

$$-a \cdot 0 = 0.$$

Then apply additive inverse to get

$$-a(b + (-b)) = 0,$$

and by the distributive law, we have

$$(-a)b + (-a)(-b) = 0.$$

By example 6 we have

$$-ab + (-a)(-b) = 0.$$

By closure we can add ab to both sides

$$ab + (-ab) + (-a)(-b) = ab + 0.$$

By additive inverse we have

$$0 + (-a)(-b) = ab + 0.$$

By commutativity, we have

$$(-a)(-b) + 0 = ab + 0,$$

and finally by additive inverse we have that

$$(-a)(-b) = ab \quad \square$$

Example 8   If $ab = 0$ then $a = 0$ or $b = 0$.

Proof   We start with
$$ab = 0,$$
then if $b \neq 0$ by multiplicative inverse there exists b^{-1} such that
$$abb^{-1} = 0b^{-1}$$
$$a \cdot 1 = 0b^{-1}$$
and
$$a = 0b^{-1}.$$
Then by commutativity of multiplication, we have
$$a = b^{-1}0,$$
and from example 5 we have
$$a = 0.$$
On the other hand, if $a \neq 0$ then from
$$ab = 0$$
we multiply by a^{-1} to get
$$a^{-1}ab = a^{-1}0,$$
and then by commutativity
$$aa^{-1}b = a^{-1}0.$$
Next by multiplicative inverse we have
$$1 \cdot b = a^{-1}0,$$
$$b = a^{-1}0,$$
and finally by example 5 we have
$$b = 0 \,{}_\square$$

Note that this also proves the contrapositive of the original statement which is: if neither $a = 0$ nor $b = 0$ then $ab \neq 0$.

Cancellation law for multiplication   If $ac = bc$ and $c \neq 0$ then $a = b$.

Proof   Start with the given:

$$ac = bc,$$

and apply the multiplicative inverse and associativity to get

$$a(cc^{-1}) = b(cc^{-1})$$

$$a1 = b1.$$

Then by multiplicative identity we have

$$a = b \;_\square$$

Example 9   Prove for $a \neq 0$ that $(a^{-1})^{-1} = a$.

Proof   Since a^{-1} is rational, by multiplicative inverse we have that

$$a^{-1}(a^{-1})^{-1} = 1.$$

Multiply both sides by a

$$aa^{-1}(a^{-1})^{-1} = a1,$$

and apply the multiplicative inverse to aa^{-1} on the left and multiplicative identity to the right side to get

$$1(a^{-1})^{-1} = a.$$

Then apply the commutative property and multiplicative identity to the left side to get the result

$$(a^{-1})^{-1} = a \;_\square$$

With these last examples you see how the usual algebra operations are combined with the equality relation and result from the field axioms. In the next group of examples we explore how the usual algebra operations are combined with the order relations.

1.2.2   Order Properties of $\mathbb{Q}$

We naturally use order in everyday life. We might put our money in order by the number on the bills before we count it. We stand in a line for coffee or fast food and observe that the line is short or long based on the number of people in the line. So our intuition is very well developed, but we probably can't immediately say if $\frac{235}{1057}$ is more or less than $\frac{272}{1147}$, but we do know that we can count things or parts of things and we do this with positive rational numbers, $\mathbb{Q}_+$. Then for any two numbers we know one is greater than the other (or they are equal). We call this trichotomy because exactly one of three possibilities must be true. Another familiar notion is that of area. For a square we multiply the lengths of the sides (two positive numbers) and get a positive number as a result. Similarly if we add the areas of two squares, we also get a positive number as a result. This familiar idea is naturally called positivity. The two ideas are stated as axioms below.

P1 *Positivity*: If $a, b \in \mathbb{Q}_+$ then $ab \in \mathbb{Q}_+$ and $(a+b) \in \mathbb{Q}_+$.

P2 *Trichotomy*: for any two numbers $a, b \in \mathbb{Q}$ exactly one of the following is true:

$$a > b$$
$$a = b$$
$$a < b$$

For this to be precise we need a definition for the inequality relation.

Inequality   For $a, b \in \mathbb{Q}$, $a > b$ is defined to mean $(a-b) \in \mathbb{Q}_+$. Similarly $a < b$ is defined to mean $(b-a) \in \mathbb{Q}_+$.

Note that the strict inequalities above are the negation of the non-strict inequalities $\leq$ and $\geq$. So for example $a \geq b$ means a is not less than b, and $a \leq b$ means a is not greater than b. In the following examples we show how the order properties, as represented by inequalities, are derived from the two axioms above.

Antisymmetry of inequality   Prove for $a, b \in \mathbb{Q}$ if $a \leq b$ and $b \leq a$ then $a = b$.

Proof   Since $a \leq b$ we know a is not greater than b. Since $b \leq a$ we know b is not greater than a. Then by trichotomy the only remaining option is that $a = b$ $_\square$

This is one way to show that two things are equal, especially when we don't have a given equality to start with.

Transitivity of inequality   Prove for all $a, b, c \in \mathbb{Q}$ if $a < b$ and $b < c$ then $a < c$.

Proof   Since $a < b$ we have that $(b - a) \in \mathbb{Q}_+$ and since $b < c$ we have that $(c - b) \in \mathbb{Q}_+$. Then by positivity we have that

$$(b-a)+(c-b)\in \mathbb{Q}_+$$

By commutativity and associativity of addition

$$(b-b)+(c-a)\in \mathbb{Q}_+$$

By the additive inverse and additive identity properties

$$(c-a)\in \mathbb{Q}_+$$

Then by definition of the less than sign we have that $a < c$ $_\square$

This allows us to make strings of inequalities like we used in example one.

Sets of Numbers   These order properties make it easy to describe sets of numbers in interval notation. We could have a set of numbers listed explicitly as in

$$\{4, 5, 3, 0, 2, 1, 3/4\}.$$

Or if we want all the numbers between two of them (excluding end points) we use the interval so that

$$(a, b) = \{x \mid a < x < b\}.$$

Alternatively, when we want to include endpoints we use square brackets:

$$[a,b] = \{x \mid a \le x \le b\}.$$

Example 10   Prove if $a > 0$ then $-a < 0$.

Proof   By definition of inequality

$$(a - 0) \in \mathbb{Q}_+$$

By the additive identity

$$a \in \mathbb{Q}_+$$

and by example 4 and the existence of the additive identity we have that

$$0 - (-a) \in \mathbb{Q}_+$$

Therefore $-a < 0$ $_\square$

Example 11   Prove if $a > 0$ and $b < c$ then $ab < ac$.

Proof   Since $a > 0$ we have that

$$(a - 0) \in \mathbb{Q}_+$$

and since $b < c$

$$(c - b) \in \mathbb{Q}_+$$

therefore by positivity their product is positive also, that is

$$(a - 0)(c - b) \in \mathbb{Q}_+$$

Then by $Q7, Q9$, and distributivity we have

$$(ac - ab) \in \mathbb{Q}_+$$

Therefore $ac > ab$ $_\square$

Example 12 Prove if $a < 0$ and $b < c$ then $ab > ac$.

Proof Since $a < 0$ we have that
$$(0-a) \in \mathbb{Q}_+$$
and since $b < c$
$$(c-b) \in \mathbb{Q}_+$$
therefore by positivity their product is positive also, that is
$$(0-a)(c-b) \in \mathbb{Q}_+$$
Then by $Q5, Q7, Q9$, and distributivity we have
$$(ab-ac) \in \mathbb{Q}_+$$
Therefore $ab > ac$ □

Example 13 Prove if $a \neq 0$ then $aa > 0$.

Proof By trichotomy either $a > 0$ or $a < 0$. If $a > 0$ then
$$(a-0) \in \mathbb{Q}_+$$
and by positivity
$$(a-0)(a-0) \in \mathbb{Q}_+$$
then applying $Q5, Q7, Q9$ twice yields
$$aa \in \mathbb{Q}_+$$
For the other case, $a < 0$ we have that
$$(0-a) \in \mathbb{Q}_+$$
and by positivity
$$(0-a)(0-a) \in \mathbb{Q}_+$$
then $Q5, 7, 9$ yields
$$(-a)(-a) \in \mathbb{Q}_+$$
and finally apply example 7 to see that
$$aa \in \mathbb{Q}_+$$
Therefore when $a \neq 0$ we have that $aa > 0$ □

Example 14   Prove if $0 < a < b$ then $0 < b^{-1} < a^{-1}$.

Proof   By transitivity we have that $0 < b$ which means

$$(b-0) \in \mathbb{Q}_+$$

and that b is positive. The motivation for the next step is that we want to show that b^{-1} is positive.

Recall example 13 that says for any $a \in \mathbb{Q}$ we have that $aa > 0$. Apply this to b^{-1} to see that $b^{-1}b^{-1} > 0$ so that

$$(b^{-1}b^{-1} - 0) \in \mathbb{Q}_+$$

By positivity, multiplying by b results in another positive number

$$b(b^{-1}b^{-1} - 0) \in \mathbb{Q}_+$$

then by distributivity

$$bb^{-1}b^{-1} - b0 \in \mathbb{Q}_+$$

and by multiplicative identity and example 5 we have

$$(b^{-1} - 0) \in \mathbb{Q}_+$$

which can be rewritten as $0 < b^{-1}$. Next we use the second given inequality, $a < b$ which means

$$(b-a) \in \mathbb{Q}_+$$

Then since a^{-1} is positive, by positivity we have

$$a^{-1}(b-a) \in \mathbb{Q}_+$$

then by distributivity and multiplicative identity we have

$$(a^{-1}b - 1) \in \mathbb{Q}_+$$

Now multiply on the right by b^{-1} and apply the distributive law to get

$$(a^{-1}bb^{-1} - 1b^{-1}) \in \mathbb{Q}_+$$

Then by the multiplicative identity and inverse rules we get

$$(a^{-1} - b^{-1}) \in \mathbb{Q}_+$$

which means that $b^{-1} < a^{-1}$ and when combined with $0 < b^{-1}$ from above we have the desired result that $0 < b^{-1} < a^{-1}$ □

Example 15   Prove if $a < b < c < d$ then $d - a > c - b$

Proof   From the given inequality we have that

$$(d-c) \in \mathbb{Q}_+$$

and that

$$(b-a) \in \mathbb{Q}_+$$

Then by positivity we have that

$$((d-c)+(b-a)) \in \mathbb{Q}_+$$

By associativity and commutativity we have that

$$(d-a-c+b) \in \mathbb{Q}_+$$

Then by distributivity with -1 we have that

$$((d-a)-(c-b)) \in \mathbb{Q}_+$$

Therefore $d - a > c - b$ $\square$

Comment   The previous examples illustrate how all the usual rules of algebra and inequalities can be built up from the field axioms, positivity, and trichotomy. From now on we will proceed as in the first three examples where we use the rules of algebra without any formal justification.

Square root question   If $b \in \mathbb{Q}$ and $rr = b$ is $r \in \mathbb{Q}$?

Discussion   This is left initially for the reader to think about. The trouble with this statement is that all the axioms we have so far have to do with rational numbers and natural numbers. We need some more structure in order to decide if something is a rational number or not. This will come with the concept of divisibility, but we'll introduce the function as well.

1.3 Functions

A function is commonly defined as a rule that maps or relates elements of one set to elements of another set. The first set is called the domain, and the second is called the range. Functions can be presented as a graph, a table of values, or as an algebraic expression. We'll almost always use algebraic expressions because they are more precise than a graph and can accommodate more possible values than a table. However, it is useful to have a precise definition for a function, so here it is.

Function   A function, f, is a set of ordered pairs that satisfy the condition

$$(x, y) \in f \text{ and } (x, z) \in f \rightarrow y = z.$$

The Domain of f is the set of all possible x values, X.
The Range of f is the set of all possible y values, Y.
The commonly used notation for a function is

$$f : X \rightarrow Y \text{ given by the rule } f(x) = \text{ algebraic expression of } x$$

Example 16 $s : \mathbb{N} \rightarrow \mathbb{N}$ given by the rule $s(n) = n + 1$ is the successor function given as an axiom of the integers.

Example 17 $f : [-1, 1] \rightarrow [-1, 1]$ given by the rule $f(x) = \pm\sqrt{1 - x^2}$ is not a function. Observe that $(0, 1) \in f$ and that $(0, -1) \in f$ but that $1 \neq -1$ so that the required condition for a function is not met. In words we say that a function must be single valued.

Inverse Function   Let $f : D \rightarrow R$ be such that $f(x) = y$. If for each y in Y, a subset of R, there is only one x with $f(x) = y$ we say f has an inverse on Y, and $f^{-1}(y) = x$.

In the rest of this short section we define the usual algebraic notation for exponents, division, and summation. Surprisingly, we can develop a strong foundation for analysis without saying anything about properties such as injective, surjective, and bijective for functions. First we'll define the term graph to be a set of ordered pairs that we usually visualize on a plane. So while the relation in example 17 is not a function, we'll call it a graph for convenience.

Graph   A graph is a set of points in a plane designated by ordered pairs, (x, y).

Example 18   So the rule given in example 17, $y = \pm\sqrt{1 - x^2}$ which might also be written as $x^2 + y^2 = 1$, has a circle centered at the origin as its graph.

Note that this notion of graph is rather specialized for visualizing relations that are easily drawn. There is a whole branch of mathematics for the study of graphs that are composed of vertices and edges that is not discussed in this book.

The following comments and examples make the the usual notation used when writing algebraic expressions a little more precise.

- An positive integer exponent indicates the number of times one number is multiplied. So for example

$$aaaaa = a^5$$

- The multiplicative inverse of a number, a^{-1} can be written as $\frac{1}{a}$.

- Then if the multiplicative inverse is multiplied several times this is notated with a negative exponent, for example

$$a^{-1}a^{-1}a^{-1}a^{-1}a^{-1} = a^{-5}$$

- The sum of k numbers can be written in summation notation like this.

$$\sum_{k=1}^{7} k = 1 + 2 + 3 + 4 + 5 + 6 + 7$$

or use the index, k to identify some unknown quantities as in

$$\sum_{k=1}^{7} x_k = x_1 + x_2 + x_3 + x_4 + x_5 + x_6 + x_7$$

It will be very useful to indicate the upper limit of the sum with a parameter such as n, like this:

$$\sum_{k=1}^{n} x_k = x_1 + x_2 + x_3 + \cdots + x_n$$

Or, if we have a formula to compute each term in the sum, like this:

$$\sum_{k=1}^{n} \frac{x^k}{k!} = \frac{x}{1} + \frac{x^2}{2} + \frac{x^3}{6} + \cdots + \frac{x^n}{n!}$$

where $k! = k(k-1)(k-2)\cdots 1$ and 0! is defined to be 1.

- The nth-root function returns the number which when multiplied by itself n times yields the original number. So for example

$$\sqrt[5]{32} = (32)^{\frac{1}{5}} = \sqrt[5]{2\cdot 2\cdot 2\cdot 2\cdot 2} = 2$$

- The so called 'power' or 'exponent' rules are just a shorter way of writing a product of many numbers (some of which may be inverses or roots as described above). Since we have justified the four basic operations on rational numbers (in section 1.2) we should apply these notational rules freely.

$$a^c a^d = a^{c+d}$$

- A couple standard conventions we follow are that for any $a \in \mathbb{Q}$:

$$a^0 = 1, \quad 0^a = 0, \quad \text{and} \quad 0^0 = 1$$

However, there is a difficulty that arises. The nth-root notation actually applies a function to a rational number and we have no guarantee from the axioms that what results will be a rational number.

The trouble with writing things in a new notation is that what we write might not make sense in terms of the objects we are applying the notation to (the natural numbers or the rational numbers). A famous example is that $\sqrt{2}$ is not rational. This is proven so many times in so many places that we won't repeat the proof by contradiction here. Instead we will introduce the concept of divisibility and prove the rational zeroes theorem as a tool to decide if the root of a polynomial is rational or not.

1.4 Divisibility

This topic would normally not be covered in an analysis book, but the concept is crucial to understanding the rational zeroes theorem. Several other books prove that $\sqrt{2}$ is irrational using proof by contradiction, but they leave the reader wondering if the same technique is necessary (or appropriate) for any other irrational numbers. We provide the rational zeroes theorem to have a tool to decide if the roots of a polynomial are rational or not. So first we define what it means for one natural number to divide another. We use natural numbers so we're not listing all the negatives (until they are needed for a further computation in an example).

Divides, Multiple, Factor   Let $m, n \in \mathbb{N}$ and $m \neq 0$. Then if there exists $p \in \mathbb{N}$ so that

$$mp = n$$

we say that m divides n or that n is a multiple of m. Also m and p are factors of n.

And symbolically write this as $m \mid n$. So for example, $3 \mid 12$ but $5 \nmid 12$. Next we define a polynomial.

Polynomial   A function $f(x) : X \to Y$ of the form

$$f(x) = \sum_{k=0}^{n} a_k x^k = a_n x^n + a_{n-1} x^{n-1} + \cdots + a_0$$

where $a_k \in \mathbb{Z}$ (for our initial purposes) and $x \in X$ is a polynomial.

Now we can observe that when we write $\sqrt{2}$ what we precisely mean is the root of the polynomial $f(x) = x^2 - 2$, that is, the x value that satisfies $x^2 - 2 = 0$. So the 5th root of 32 from above is actually the root of the polynomial $g(x) = x^5 - 32$. With the precise meaning of terms established we'll move on to state, discuss, and prove the rational zeroes theorem. In what follows p, q, n, k are all integers and we restrict any denominator to not be zero while x is any number.

Rational Zeroes Theorem   Let $f(x): X \rightarrow Y$ be a polynomial:

$$f(x) = \sum_{k=0}^{n} a_k x^k.$$

Then any rational number $x = \frac{p}{q}$ such that $f(x) = 0$ also has q as a factor of $a_n \neq 0$ and p as a factor of $a_0 \neq 0$.

Discussion   Usually the first choice for proving a statement is to use direct proof. We have some given information we know we can use, and the conclusion. We should write these things in mathematical notation and then apply the rules (axioms, definitions, and other theorems) to try to link the given to the conclusion. So we have a polynomial of a number designated as $x = \frac{p}{q}$ which we can write as

$$a_n \left(\frac{p}{q}\right)^n + a_{n-1}\left(\frac{p}{q}\right)^{n-1} + \cdots + a_0 = a_n p^n q^{-n} + \cdots + a_1 p q^{-1} + a_0 = 0.$$

We also know that if $mp = n$ then m and p are factors of n. So we need to use the notion of factor on a_n and a_0 (from the conclusion of the theorem). In order to do this we start to solve for them from the equality on the polynomial. As we do the algebraic manipulation we look for the opportunity to apply the definition of factor:

$$a_n p^n q^{-n} + a_{n-1} p^{n-1} q^{1-n} + \cdots + a_1 p q^{-1} = -a_0.$$

Note that we can factor out the p here:

$$p\left(a_n p^{n-1} q^{-n} + a_{n-1} p^{n-2} q^{1-n} + \cdots + a_1 q^{-1}\right) = -a_0.$$

So p is a factor of $-a_0$. And if we isolate the a_n term instead we get

$$a_{n-1} p^{n-1} q^{1-n} + \cdots + a_1 p q^{-1} + a_0 = -a_n p^n q^{-n}$$

which is the same as

$$q\left(a_{n-1} p^{n-1} + \cdots + a_1 p q^{n-2} + a_0 q^{n-1}\right) = -a_n p^n,$$

so that q is a factor of $a_n p^n$ which at first doesn't look promising. But we usually think of rational numbers in lowest terms; we say 3/4 and not 12/16. So if p and q have no common factors, then q is a factor of a_n. With this insight the first line of the proof makes more sense.

Rational Zeroes Theorem   Let $f(x): X \to Y$ be a polynomial:

$$f(x) = \sum_{k=0}^{n} a_k x^k.$$

Then any rational number $x = \frac{p}{q}$ such that $f(x) = 0$ also has q as a factor of $a_n \neq 0$ and p as a factor of $a_0 \neq 0$,

Proof   Let $x = \frac{p}{q}$ be in lowest terms (p and q have no common factor) and since $f(x) = 0$ we have that:

$$a_n p^n q^{-n} + a_{n-1} p^{n-1} q^{1-n} + \cdots + a_1 p q^{-1} + a_0 = 0$$

and so

$$p\left(a_n p^{n-1} q^{-n} + a_{n-1} p^{n-2} q^{1-n} + \cdots + a_1 q^{-1}\right) = -a_0.$$

Hence p is a factor of a_0. Next isolate the first term on the right side.

$$a_{n-1} p^{n-1} q^{1-n} + \cdots + a_1 p q^{-1} + a_0 = -a_n p^n q^{-n}$$

Multiply by q^n to get

$$q\left(a_{n-1} p^{n-1} + \cdots + a_1 p q^{n-2} + a_0 q^{n-1}\right) = -a_n p^n.$$

And since q and p have no factors in common, nor does q and p^n. Therefore q is a factor of a_n $\square$

Now we know that any rational root of a polynomial with $a_k \in \mathbb{Z}$ must be constructed from the factors of the first and last constant of the polynomial (when it is written in the usual order of decreasing exponents). This says that the only rational numbers that could be the solution of $x^2 = 2$ have $p = \pm 2$ and $q = \pm 1$ which is precisely 2 and −2. These are obviously not the roots ($2^2 = 4$), and since there are no other possibilities, the roots must not be rational. This is not an isolated case, consider the following equalities where similar logic applies (maybe write some of your own in the white space):

$$x^2 = 3$$
$$x^3 = 4$$
$$x^3 = 5$$
$$x^4 = 6$$

So what do we do now? With just a few axioms and definitions, and one theorem, we discovered something new. When this happens in math we add another definition to account for this new thing. But this thing is so important that we have another axiom to account for it. Notice that in the definition of polynomial the variable x is completely arbitrary. It is not necessarily any type of number that we've specified.

So we will add the necessary axiom in the next section. The new set of numbers is the completion of the rational numbers, called the real numbers. After we work with sequences in chapter two we'll define what it means to be complete in precise terms. For now we'll just add the axiom and call it the completeness axiom.

Another direction this book could take would be to consider the roots of the polynomial $x^2 + 1 = 0$. If we did, the subject would be complex variables. Alas, maybe that will come next, but we have quite a way to go with real variables.

The Completeness Axiom, along with the Archimedean Property, and the density of the rationals in the real numbers give us the essential properties of numbers that we take for granted, so we'll cover these topics next.

1.5 Real Number Field

Let us consider what we need from this new axiom in order to generate it. In simple language we have discovered some holes in the rational numbers that are roots of polynomial functions. Since our rational numbers are ordered, it is natural to think of them as arranged along a line from smaller on the left to bigger on the right. We could collect up a section of them and go about finding the holes, but it turns out that there are too many holes to count, in the common meaning of 'count'. Instead we can just look at one end of an arbitrary collection from the line, and construct our axiom so that it says there is always an end point at the big end of the collection, regardless of the collection we have. This collection of numbers is called a set, and we have assumed you are familiar with sets. The following ideas spell out some details:

- If there is going to be an end point of the set, the set must be bounded.

  Bounded A set of real numbers, $X \subset \mathbb{R}$ is *bounded above* if there exists $M \in \mathbb{R}$ such that for all $x \in X$, $x \leq M$. Similarly, a set of real numbers, $X \subset \mathbb{R}$ is *bounded below* if there exists $m \in \mathbb{R}$ such that for all $x \in X$, $x \geq m$.

- We want every bounded set of numbers to have end points.

  Supremum Let X be a subset of real numbers that is bounded above. The supremum of X is the least upper bound.

  Infimum Let X be a subset of real numbers that is bounded below. The infimum of X is the greatest lower bound.

- We'll collect a set of real numbers in such a way that we only have to examine the largest number of (but not necessarily in) the set.

Completeness Axiom Every non-empty set, X, of real numbers, bounded above, has a least upper bound, called the supremum of X, or sup X, and sup $X \in \mathbb{R}$.

Now that we have added the Completeness Axiom, we need to update, or at least re-interpret the field axioms we previously used on the rational numbers. They become the axioms of the real numbers by replacing $\mathbb{Q}$ with $\mathbb{R}$ and adding the completeness axiom. The results of all the previously justified algebraic manipulations remain valid for the real numbers as well.

The Field Axioms for Real Numbers, $\mathbb{R}$.

R1 If $a, b \in \mathbb{R}$ then $(a + b) \in \mathbb{R}$.

R2 If $a, b \in \mathbb{R}$ then $(ab) \in \mathbb{R}$.

R3 If $a, b, c \in \mathbb{R}$ then $(a + b) + c = a + (b + c)$.

R4 If $a, b, c \in \mathbb{R}$ then $(ab)c = a(bc)$.

R5 If $a, b \in \mathbb{R}$ then $a + b = b + a$.

R6 If $a, b \in \mathbb{R}$ then $ab = ba$.

R7 There exists $0 \in \mathbb{R}$ such that for all $x \in \mathbb{R}$, $x + 0 = x$.

R8 There exists $1 \in \mathbb{R}$ such that for all $x \in \mathbb{R}$, $x \cdot 1 = x$.

R9 For all $a \in \mathbb{R}$ there exists $-a \in \mathbb{R}$ such that $a + (-a) = 0$.

R10 For all $a \in \mathbb{R} \setminus \{0\}$ there exists $a^{-1} \in \mathbb{R}$ such that $a(a^{-1}) = 1$.

R11 If $a, b, c \in \mathbb{R}$ then $a(b + c) = ab + ac$.

R12 Every non-empty set, X, of real numbers, bounded above, has a least upper bound, called the supremum of X, or $\sup X$, and $\sup X \in \mathbb{R}$.

Before we show how this updated system plugs the hole where $\sqrt{2}$ sits, we'll establish the Archimedean Property which says there is always an integer larger than a particular real number.

This page intentionaly left blank.

Archimedean property   Let $x \in \mathbb{R}$; then there exists $n \in \mathbb{N}$ such that $n > x$.

Discussion   This example and the next one continue the use of sets of real numbers. By proving the Archimedean property and the density of the rationals in the real numbers we justify that choosing a rational number will be 'good enough.'

Now how do we decide what kind of set we need to use? If we choose a set with natural numbers greater than x, we are using the idea that we want to prove, so we need to choose a set with natural numbers less than x, like this:

$$S = \{n \in \mathbb{N} \mid n \leq x\}.$$

Then any set is empty or not, making two cases. Since we need a conclusion that has to do with inequalities, we can apply trichotomy, and this will lead to the desired conclusion if the set is empty.

If the set is not empty, we observe that the natural numbers are a subset of the real numbers and use the supremum of the set, s. Then there ought to be something in the set that is less that s, and we want it to be a natural number. Nothing tells us that the supremum of a set of natural numbers is a natural number. To get a natural number we have to get back in the set (since the set is of natural numbers).

The insight that will get us back inside the set is the idea of not being an upper bound. If s is the least upper bound, then $s - 1$ is not an upper bound. What does this mean? Recall the meaning of an upper bound of a set which says there exists $M \in \mathbb{R}$ with M greater than or equal to anything in the set. Then, if a number, $s - 1$, is not an upper bound, there must be an element in the set greater than $s - 1$. Since our set only has natural numbers, the element will be a natural number, say m, and

$$m > s - 1.$$

Then by the successor operation we have

$$m + 1 > s,$$

and by the definition of supremum we get the desired result for the case when S is not empty as well.

Archimedean property Let $x \in \mathbb{R}$; then there exists $n \in \mathbb{N}$ such that $n > x$.

Proof Consider the following set:

$$S = \{n \in \mathbb{N} \mid n \leq x\}.$$

It is empty or not, so we have two cases:

C1 For $S = \varnothing$, by trichotomy we have either $0 \leq x$ or $0 > x$. Since S is empty we cannot have $0 \leq x$ because then $0 \in S$. Therefore $0 > x$ and $0 \in \mathbb{N}$ by the first axiom of the natural numbers.

C2 For $S \neq \varnothing$, note that S is bounded above by x (by definition of S). Since S is a non-empty subset of the real numbers, bounded above, it has a supremum and

$$s = \sup S \in \mathbb{R}.$$

Therefore $s - 1$ is not an upper bound of S, and so there exists $m \in S$ such that

$$m > s - 1.$$

Since $m \in S$, $m \in \mathbb{N}$ by definition of S, and by axiom 2 of the natural numbers $m + 1 \in \mathbb{N}$ with

$$m + 1 > s.$$

Since s is the supremum, it is an upper bound and therefore greater than or equal to x so

$$m + 1 > s \geq x,$$

and the integer greater than x is $m + 1$ $\square$

Note that we can manipulate the conclusion in several ways with the rules of algebra. In each case the conclusion that there exists $n \in \mathbb{N}$ holds. Dividing by n yields

$$1 > \frac{x}{n},$$

or multiplying by $y > 0$ with $z = xy$ yields

$$ny > z.$$

Next is the density property that tells us there is always a rational number between any two distinct real numbers.

Density of Rational Numbers in Real Numbers   Suppose $a, b \in \mathbb{R}$ and $a < b$. Then there exists a rational number, $r \in \mathbb{Q}$, such that $a < r < b$.

Discussion   From $a < b$ we have that $b - a > 0$ and from the Archimedean property we can get a natural number n such that $n > b - a > 0$, but this isn't quite enough. We want

$$a < \frac{m}{n} < b$$

or

$$an < m < bn$$

which means that m is an integer, so that $bn - an > 1$. This rearranges to be

$$n > \frac{1}{b - a}.$$

Then to get m we'll construct a set of natural numbers (since we want m to be a natural number). Since we have made enough room between na and nb for one natural number we can construct the set so that the smallest natural number in the set is the one we want, that is, closest to na as follows

$$\{x \in \mathbb{N} \mid x > na\}.$$

We'll take the intuitively obvious idea that a non-empty set of natural numbers contains its smallest element as a fact, and call that smallest element m in this case. Then $m > an$, and since m is the smallest, $m - 1$ is not the smallest, and therefore not in the set, hence

$$m - 1 \leq an.$$

The remaining algebra is included in the proof on the next page. We note that when we wrote the set

$$\{x \in \mathbb{N} \mid x > na\}$$

that it is only useful when $a \geq 0$ so that a different case is required for $a < 0$. This other case is straightforward when $b > 0$ since zero is rational. When both a and b are negative we are just reflected about the origin and the negative of each number gives us the above case with a and b flipped in order.

Density of Rational Numbers in Real Numbers   Suppose $a, b \in \mathbb{R}$ and $a < b$. Then there exists a rational number, $r \in \mathbb{Q}$, such that $a < r < b$.

Proof   Note that since $b > a$ we have that $b - a > 0$, and therefore

$$\frac{1}{b-a} \in \mathbb{R}.$$

Then by the Archimedean property we have an $n \in \mathbb{N}$ such that

$$n > \frac{1}{b-a} \implies \frac{1}{n} < (b-a).$$

Now for the case when $a \geq 0$ consider the following set M:

$$M = \{x \in \mathbb{N} \mid x > an\}.$$

By the Archimedean property M is not empty. Since M is a set of natural numbers it contains it's smallest element (in some texts this statement is proven but we omit the proof here). Let the smallest element be m, and then we have that

$$m > an$$

which immediately gives us one needed inequality, $\frac{m}{n} > a$. Then continuing on from $m > an$, if m is the smallest, then $m - 1$ is not in the set and

$$m - 1 \leq an.$$

Applying algebra yields

$$m \leq an + 1$$
$$\frac{m}{n} \leq a + \frac{1}{n}.$$

Then using $\frac{1}{n} < (b-a)$ from above yields

$$\frac{m}{n} < a + (b-a) = b.$$

Then we have a rational number, $\frac{m}{n}$ such that

$$a < \frac{m}{n} < b.$$

There remains the case when $a < 0$. Then if $b > 0$ we observe that 0 is a rational number between a and b. For $b \leq 0$ we have $0 \leq -b < -a$ and the first case applies since $-b \geq 0$ $\square$

1.6   The Secant Method

We have covered two ideas that might seem to conflict. We saw examples of 'holes' in the rational numbers, and after that said that no matter how close two distinct real numbers are, that there is a rational number between them. What this means is that the rational numbers are good enough to compute whatever it is that you need to compute. One popular method for computing the root of an equation is the secant method. The secant method uses a line to approximate the function and so to get started it needs two points. Recall the point slope form of the equation of a line is:

$$y - y_0 = m(x - x_0).$$

We'll find a root of the equation $x^2 - 3 = 0$ by starting with the points $(0, -3)$ and $(2, 1)$ which are on the curve $x^2 - 3$. The picture on the next page illustrates the process graphically. First, $l_1(x)$ is constructed as a line between the two points:

$$y - (-3) = \frac{4}{2}(x - 0)$$

$$y = 2x - 3 = l_1(x).$$

Then since we want the root, we know $y = 0$ and therefore we solve for x to get $x = 1.5$. This is shown by the vertical line between the original, solid function, and the dashed line of our first approximation. Now we have that $f(1.5) = -.75$ and go on to make the next approximation represented by the dotted line, $l_2(x)$. If we specify the process using subscripts then our initial guesses are labeled x_{n-1} and x_n. To compute the next iterate we use the previous two guesses as follows:

$$x_{n+1} = x_n - \frac{f(x_n)(x_n - x_{n-1})}{f(x_n) - f(x_{n-1})}.$$

In this example the initial guesses were $x_0 = 0$ and $x_1 = 2$. When we use a computer to do several iterations we can generate a table of the results like this:

Iterate	Initial, then Computed Root
0	0.00000000
1	2.00000000
2	1.50000000
3	1.71428571
4	1.73333333
5	1.73204420
6	1.73205081
7	1.73205081

And so it appears that the list of numbers is getting closer and closer to the number that is the root of the equation. But what does this mean precisely? First we'll define absolute value and show that it is a metric, something to use to measure how close two mathematical objects (like numbers) are to each other. Then we'll define a list of numbers to be a sequence which is a special type of function. Before we do all that, lets take a look at some more interesting applications of the secant method. We'll see what can go wrong, and why we need to be precise.

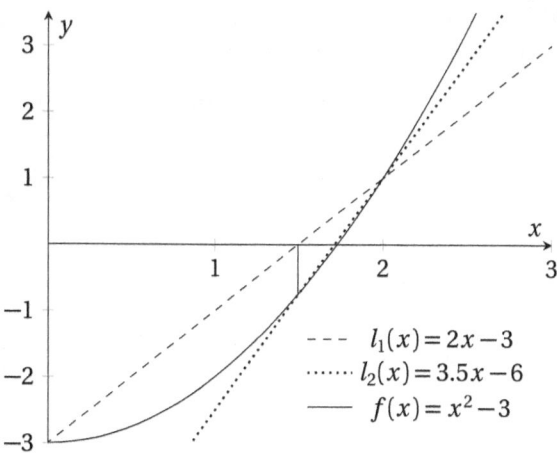

$$l_1(x) = 2x - 3$$
$$l_2(x) = 3.5x - 6$$
$$f(x) = x^2 - 3$$

Consider the expression

$$\sqrt{x+1}-\sqrt[3]{x}-.5=0$$

and try to solve for x. You will probably encounter some difficulty, but now you have the secant method to use. Depending on your initial guess for the solution, the secant method results in a table like this:

Iterate	Initial guess 2,1	Initial guess 1/2, 0
0	2.00000000	0.50000000
1	1.00000000	0.00000000
2	2.48121673	0.43940156
3	2.29943807	0.39197642
4	2.34014699	0.10063614
5	2.34056610	0.28052353
6	2.34056475	0.24182119
7	2.34056475	0.21854731
8	2.34056475	0.22204512
9	2.34056475	0.22186210
10	2.34056475	0.22186039
11	2.34056475	0.22186039

It appears to be working, especially when we plot what the function looks like below.

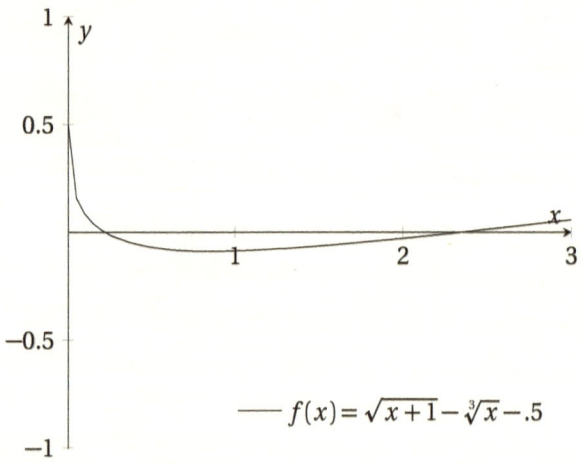

$$f(x)=\sqrt{x+1}-\sqrt[3]{x}-.5$$

Notice that the root we find depends on the initial guess we supply to the procedure. If our guess is too far off we might not find a root. For the function above, initial guesses of $x_0 = 14$ and $x_1 = 15$ can result in an error from your computer. Also, observe what happens when we apply the procedure to the expression

$$x^2 + 1 = 0.$$

Iterate	Initial guess 0, 1/2	Initial guess 14, 20
0	0.00000000	14.00000000
1	0.50000000	20.00000000
2	-2	8.205882353
3	1.333333333	5.783107404
4	5.5	3.320861606
5	0.926829268	1.999666225
6	0.637571157	1.060160753
7	-0.261493403	0.366023196
8	-3.102338909	-0.429086706
9	0.056114225	18.34746739
10	0.385423094	-0.495170546
11	-2.215831462	-0.564920331
12	1.012906546	0.679439949
13	2.697117791	-12.08377626
14	0.466823963	0.807605114
15	0.081884951	0.954128787
16	-1.752794818	-0.130235736
17	0.684374149	-1.364572336
18	2.058709203	0.550093308
19	0.149075805	2.149401089
20	-0.313932862	0.067557523
21	6.34974209	-0.385569573
22	-0.49593892	3.22644417
23	-0.708784377	-0.789904174
24	0.538286462	-1.456401998
25	8.102908647	-0.066962384

Since this method isn't working, we need to develop some theory to tell us why.

1.7 Absolute Value as a Metric

The previous section on the secant method was included to generate some interest and curiosity. Hopefully the examples presented there revealed some limits on our rules of algebra, that we can write math expressions that we can't solve with the usual algebra rules. But, while we make up some of that deficiency with procedures like the secant method, they do not work in all cases. Now back to some theory, because, "With theory, we can separate fundamental characteristics from fascinating idiosyncrasies and incidental features."

1.7.1 Absolute Value

Absolute value is a function from the set of real numbers to the set of non-negative real numbers.

Absolute Value: For $x \in \mathbb{R}$, if $x \geq 0$ then $|x| = x$. If $x < 0$ then $|x| = -x$.

We should interpret absolute value as the distance between a real number and zero since the additive identity property tells us that $|x - 0| = |x|$. Then if we generalize, $|x - y|$ is the distance between x and y. The next two proofs are a direct application of the definition. Following them are the triangle and reverse triangle inequalities which are crucial to all of analysis.

Example 19 For $x \in \mathbb{R}$, $|x| \geq 0$.

Proof We'll consider the three cases from the trichotomy property:
Case1: $x = 0$. Then by definition, $|0| = 0$ and $0 \geq 0$.
Case2: $x > 0$. Then by definition, $|x| = x$ and $x \geq 0$.
Case3: $x < 0$. Then by definition, $|x| = -x$. But since $x < 0$, by the inequality rules we have $-x > 0$ and then $|x| > 0$ $\square$

This next example should be used routinely without too much thought. Note that the example above only works with non-strict inequalities but the statement below works with strict ($<,>$) inequalities.

Example 20   Let $a, x \in \mathbb{R}$ with $a \geq 0$. Then $|x| \leq a$ if and only if $-a \leq x \leq a$.

Proof   First suppose $|x| \leq a$. Then by definition of absolute value $x \leq a$, and $-x \leq a$ can be rewritten as $x \geq -a$. Apply transitivity to the first and last inequalities in the last sentence to get $-a \leq x \leq a$.
Now suppose that $-a \leq x \leq a$. Then $-a \leq x$ can be rewritten as $a \geq -x$. Since we have $x \leq a$ and $-x \leq a$ we have $|x| \leq a$ by definition of absolute value $\square$

Triangle Inequality   For all $x, y \in \mathbb{R}$ we have that

$$|x + y| \leq |x| + |y|.$$

Proof   Consider that $x \leq |x|$ and so by example 22 we have that $-|x| \leq x \leq |x|$. Similarly we have $-|y| \leq y \leq |y|$. Then by positivity we can add inequalities to get

$$-\left(|x| + |y|\right) \leq x + y \leq |x| + |y|$$

which by example 22 is equivalent to

$$|x + y| \leq |x| + |y| \square$$

Reverse Triangle Inequality   For all $x, y \in \mathbb{R}$ we have that

$$|x| - |y| \leq |x + y|.$$

Proof   This is another place where we do the add zero trick. You will see it many times in analysis. Note that $0 = y + (-y)$ so that an application of the triangle inequality yields

$$|x| = |x + y + (-y)| \leq |x + y| + |y|$$

$$|x| - |y| \leq |x + y| \square$$

Note that there is no restriction on x or y so that if we negate y the conclusion is

$$|x| - |y| \leq |x - y|$$

which might be useful as well.

1.7.2 Metric Space axioms

A first course in single variable analysis can be given without any reference to metric spaces. The machinery of absolute value is all that is needed, but then a door remains closed to the hallway of metric spaces that leads to topology. So we will open the door and take a peek down the hallway, and we will walk down the hall in the chapter on continuity. This means we'll give the properties of a metric space, and show that absolute value is a metric. Then we'll point out that there are several different ways to measure distance with some examples of different metrics.

A metric space is a set of elements (that often are not numbers, but they will be made of numbers in our examples) and a function that maps pairs of elements to the real numbers in a general way.

Metric Space A metric space is a pair (X, d) where X is a set and $d : X \times X \to \mathbb{R}$ is a function that maps any two elements from X to a real number, with the following properties:

- The mapping is *non-negative*: for all distinct $x, y \in X$ we have $d(x, y) > 0$, and $d(x, x) = 0$.

- The mapping is *symmetric*: $d(x, y) = d(y, x)$.

- The *triangle inequality* holds: for all $x, y, z \in X$ we have that $d(x, z) \le d(x, y) + d(y, z)$.

Absolute Value is a Metric Since absolute value is defined on real numbers the set is $\mathbb{R}$ and all that remains is to verify the three properties by applying the definition of absolute value.

Proof First is non-negativity. Let $x, y \in \mathbb{R}$ be distinct. Then $x - y > 0$ or $-(x - y) = y - x > 0$. Combining the cases we have $|x - y| > 0$ as desired. Then $|x - x| = |0| = 0$ so that the nonnegative property is satisfied. For symmetry note that $|a| = |-a|$ and applying this yields

$$|x - y| = |-(x - y)| = |y - x|$$

so that symmetry is satisfied. Finally, note that we proved the triangle inequality property on the previous page $\square$

Discussion   It is common to use ordered pairs such as $(2, -3)$ to specify points in the plane. So a generic pair of points would be given by $p = (x_0, y_0), q = (x_1, y_1)$. The plane where these points live is called $\mathbb{R}^2$, and one way to measure the distance between them is:

$$d(p, q) = \sqrt{(x_1 - x_0)^2 + (y_1 - y_0)^2}.$$

But we could think of a taxicab driving through Manhattan and then the distance would be

$$d_T(p, q) = |x_1 - x_0| + |y_1 - y_0|.$$

Yet another way to measure the distance is the maximum metric:

$$d_M(p, q) = \max\{|x_1 - x_0|, |y_1 - y_0|\}.$$

All three of these are metrics, but we omit the proofs here. The thing to realize here is that on the one dimension real line we have the order properties (trichotomy). In the plane or higher dimension spaces there isn't a natural ordering. So we would use a metric, that maps the distance between points back to the real line, so that we can 'order' the points.

The details to work out in metric spaces can be a bit more challenging, so that material is usually covered in a second course following this single variable version. Next we'll introduce the greek letter ϵ and it's central meaning in analysis. Then use it in a proof to show that the square root function actually works.

From the completeness axiom, we know that any value we can write as the supremum of a non-empty set exists as a real number. So if $b < a$, then $(b - a) \in \mathbb{R}_+$ and by choosing ever larger integers n, we can make

$$\epsilon = \frac{b - a}{n}$$

as small as we like. To see this, fix $N \in \mathbb{N}$ and by the completeness axiom ϵ exists as

$$\epsilon = \sup\left\{ x \,\middle|\, x < \frac{b - a}{N} \right\}.$$

Note that $\frac{b-a}{N}$ could be replaced with any other expression of a number as well.

Some ideas to keep in mind when we see ϵ are the following:

- ϵ is a real (rational or irrational) number.

- ϵ is usually small, so large values may be ignored by the proof.

- ϵ is usually not allowed to be zero, but may be arbitrarily small.

- Generally, ϵ remains a fixed value in the proof. However, the logic of the proof is designed by the proof writer to be valid for any initial value of ϵ. So when the proof works for $\epsilon = 0.123456$, it should work for any smaller one as well, say $\epsilon = 6.02 X 10^{-23}$.

An epsilon argument implicitly uses ideas like these:
If $\forall \epsilon > 0$ we have $|a - b| < \epsilon$, then

$$-\epsilon < a - b < \epsilon,$$

$$b - \epsilon < a < \epsilon + b.$$

From the quantifier, for-all, ϵ is as small as we like. If it vanished we would have

$$b < a < b$$

which doesn't make sense. As a result we interpret $|a - b| < \epsilon$ to mean $a = b$ when used under the for-all quantifier.

However, in practice there is often an integer n, or another real number such as δ, that depends on a particular value of ϵ. We'll spend a lot of time in this book exploring the relationship between ϵ and n, or between ϵ and δ, in order to draw conclusions like a=b. The Archimedean Property lets us choose larger and larger integers for smaller and smaller epsilons.

Next we give a rigorous justification for the square root function, that the square root of a positive number exists as a real number. This proof uses ϵ, but without an integer, or a δ in response.

Square root Let $b \in \mathbb{R}$ with $b > 0$, and $rr = b$. Prove that $\sqrt{b} \in \mathbb{R}$

Discussion Notice that we have this new axiom about sets of real numbers. So we need to construct a set, but how? Well, we want r to be a real number, and if r is the supremum, then by completeness it will be a real number. But the set has to have real numbers in it so we'll start with $\{x \mid x \in \mathbb{R}\}$. We also have this other condition that $rr = b$. Since completeness is only for the big end of the set, we'll fill the set with real numbers less than or equal to the big end:

$$\{x \mid x^2 \le b, \; b \in \mathbb{R}, \; x \in \mathbb{R}\}$$

Next we'll need a way to say that the supremum of this set is such that $rr = b$. Trichotomy is the only tool we have to conclude equality so far, so we'll examine the three possible cases. If we're on track the inequalities will turn out to be wrong, and the equality will be true. It might be tempting to use the symbol $\sqrt{b}$ in our reasoning, but the problem is that we don't know that is a real number, nor do we have any machinery to say that b is not a real number. What we have is that r is a real number, and a supremum, so we need to start with $rr > b$ and conclude that there is a $(r-\epsilon)^2$ greater[1] than b. This will contradict the idea that r is the *least* upper bound. The algebra that inspires the choice of ϵ is

$$(r-\epsilon)^2 = r^2 - 2r\epsilon + \epsilon^2,$$

then introduce b by adding $0 = b - b$, and drop ϵ^2 to get

$$(r-\epsilon)^2 = r^2 - 2r\epsilon + \epsilon^2 > b + (r^2 - b) - 2r\epsilon,$$

and then choose ϵ so that all that remains is b on the right side of the inequality. For $rr < b$ we want instead $(r+\epsilon)$ to be in the set, and then r cannot be the supremum. Doing similar algebra we notice an ϵ^2 term which is smaller when $\epsilon < 1$ and use that to our advantage. From the below expression we choose ϵ so that the right hand side is b, and we're lucky that we have $0 < \epsilon < 1$, as is needed.

$$(r+\epsilon)^2 < r^2 + 2r\epsilon + \epsilon = r^2 + \epsilon(2r+1)$$

---

[1] $(r^2 - \epsilon)$ doesn't work as well, and we're really after something smaller than r, not r^2.

Square root   Let $b \in \mathbb{R}$ with $b > 0$, and $rr = b$. Prove that $\sqrt{b} \in \mathbb{R}$

Proof   Consider the set X, defined below

$$X = \{x \mid x^2 \le b, b \in \mathbb{R}, \ x \in \mathbb{R}\}$$

Note that $0 = 0^2 < b$ so that $0 \in X$ and X is not empty. Also note that since $b \ge x^2$ we see that b is an upper bound of X. Then by completeness the supremum is a real number we'll call r:

$$\sup X = r \in \mathbb{R}.$$

By trichotomy only one of these three cases may be true:

$$rr > b, \quad rr < b, \quad rr = b.$$

C1: For $rr > b$ let $\epsilon = \frac{r^2 - b}{2r}$ and $\epsilon > 0$ since $r^2 > b$ and $r > 0$. Then by the rules of algebra we have

$$(r - \epsilon)^2 = r^2 - 2r\epsilon + \epsilon^2.$$

Then adding $0 = b - b$ yields

$$(r - \epsilon)^2 = b + (r^2 - b) - 2r\epsilon + \epsilon^2 > b + (r^2 - b) - 2r\epsilon.$$

Then since $\epsilon = \frac{r^2 - b}{2r}$ we have

$$(r - \epsilon)^2 > b + (r^2 - b) - 2r\frac{r^2 - b}{2r},$$

and on the right side all that is left is b, so that

$$(r - \epsilon)^2 > b.$$

This contradicts that r is the *least* upper bound (since $r - \epsilon < r$), so we must conclude that $rr \le b$.

C2: For $rr < b$ let $\epsilon = \frac{b - r^2}{2r + 1}$. Note that since r is an upper bound and $0 \in X$, $r \ge 0$. Also since $rr < b$, we know $b - r^2 > 0$. Therefore $\epsilon > 0$. In what follows we'll also need that $\epsilon < 1$. Since $r = \sup X$, we know $r + 1 \notin X$. Therefore $b < (r + 1)^2$ and

$$b < r^2 + 2r + 1$$

$$\frac{b-r^2}{2r+1} = \epsilon < 1.$$

Now consider

$$(r+\epsilon)^2 = r^2 + 2r\epsilon + \epsilon^2.$$

Since $0 < \epsilon < 1$ we have that

$$(r+\epsilon)^2 < r^2 + 2r\epsilon + \epsilon = r^2 + \epsilon(2r+1).$$

Now apply the chosen value for ϵ,

$$(r+\epsilon)^2 < r^2 + \frac{b-r^2}{2r+1}(2r+1) = b.$$

Now with $(r+\epsilon)^2 < b$ we are contradicting the fact that r is the least upper bound (since $r + \epsilon > r$). Therefore we must conclude that $rr \geq b$.

C3: Since case one concluded that $rr \leq b$ and case two concluded that $rr \geq b$, by trichotomy we must conclude that $rr = b$.

From above $r \in \mathbb{R}$, and applying the root notation to $rr = b$ we see that $r = \sqrt{b}$, and therefore $\sqrt{b} \in \mathbb{R}$ $_\square$

In essence we have proven the existence of the square root function for positive real numbers. Shortly after we define continuity we'll see that the Intermediate Value Theorem is much easier. But before we move on to more examples it is worth reflecting on the ideas used in this proof.

- We are free to construct any set of numbers we wish and use it in any way we can, consistent with set theory and our axioms.

- Since there are no 'holes' in the real numbers, we can choose them freely, unless it's a proof like this one to say that a particular thing is a real number. Notice that we freely chose ϵ without justifying its existence; the completeness axiom lets us do that.

- To conclude that two numbers were equal we showed one non-strict inequality, and then the other. This is a method used throughout analysis.

Now that we have defined the square root it is much easier to apply the rational zeroes theorem to decide that a number is irrational as in the next two examples. We'll go ahead and use all other roots freely.

Examples 21   Prove that $\sqrt{9+4\sqrt{5}}$ is irrational.

Proof   Let $x = \sqrt{9+4\sqrt{5}}$ and then

$$x^2 = 9+4\sqrt{5},$$

$$x^2-9 = 4\sqrt{5},$$

$$(x^2-9)^2 = 80,$$

$$(x^2-9)^2 - 80 = 0.$$

Therefore the possible rational roots are $\pm 2, \pm 4, \pm 5, \pm 8, \pm 10, \pm 16, \pm 20$, and ± 40. You can check all the values, or after checking the first three reason that all the larger ones (in absolute value) will fail because they all evaluate to numbers larger than zero.

$$((\pm 2)^2 - 9)^2 - 80 = -55$$

$$((\pm 4)^2 - 9)^2 - 80 = -31$$

$$((\pm 5)^2 - 9)^2 - 80 = 176$$

Then by the rational roots theorem, the root $x = \sqrt{9+4\sqrt{5}}$ is not rational. Note that $\sqrt{9+4\sqrt{5}} - \sqrt{5} = 2$ and is therefore rational $\square$

Example 22   Prove that $(1+\sqrt{3})^{2/3}$ is irrational.

Proof   Let $x = (1+\sqrt{3})^{2/3}$ and then

$$x^3 = (1+\sqrt{3})^2 = 9+6\sqrt{2}+2$$

$$x^3-11 = 6\sqrt{2},$$

$$(x^3-11)^2 = 72,$$

$$(x^2-9)^2 - 72 = 0.$$

Therefore the possible rational roots are $\pm 2, \pm 3, \pm 4, \pm 6, \pm 8, \pm 9, \pm 12, \pm 18, \pm 24$ and ± 36. Observe that all the possible rational roots with absolute vaue at or below $x = 4$ are negative and all the possible rational roots with absolute value at or above $x = 6$ are positive, so by the rational roots theorem there are no rational roots. Hence $x = (1+\sqrt{3})^{2/3}$ is not rational $\square$

Chapter 2

Sequences

We're going to use sequences to study the real numbers because the real numbers are the equivalence classes of Cauchy sequences of rational numbers. If you understand that statement, then you shouldn't be wasting your time with this book. However, you probably have the idea that a sequence is a list of numbers. So we'll be precise:

Sequence A sequence is a function from the natural numbers to the real numbers: $x_n : \mathbb{N} \to \mathbb{R}$ where $x_n = $ *some rule.*

Note the subscript notation that is used, and that the rule can be explicit as in
$$x_n = \frac{1}{n} \text{ for } n \geq 1,$$
or the rule could be implicit as in
$$b_n = 1, \frac{1}{4}, \frac{1}{9}, \frac{1}{16}, \frac{1}{25}, \ldots$$

We're free to make up sequences in any way we want, but in practice they usually consist of rational numbers. There are other approaches to learning analysis that do not spend much time with sequences. Those other approaches build from topology and metric spaces. The strength of building from sequences is that the applications in numerical analysis then seem more natural, and the extension to functional analysis

53

which deals with sequences of functions is natural as well. The drawback
is that there remains a gap in understanding the topology that must be
addressed for a full understanding in more than one dimension.

Recall the examples in the section on the secant method. In two of
those examples it appeared that the list of numbers was getting closer
and closer to a particular value. We call this convergence. But in the last
example it did not appear that the numbers were approaching a particular value. We would say that the sequence does not converge, or diverges.
Consider the following:

$$a_n = 1, 0, -1, 0, 1, 0, -1, \ldots$$

Does it converge or diverge? Surely this next one diverges

$$c_n = 1, 2, 3, 4, 5, 6, 7, 8, 9, \ldots$$

2.1   Convergent Sequence Properties

In this section we give just two proofs for the convergence of a specific
sequence. We focus on general properties that will be useful later, and
this exposes us to varied techniques to use in proofs. The initial definition of convergence is relatively straightforward, but it has the drawback
that you have to know what the limit is going to be in order to apply the
definition. With specific sequences this isn't a big deal. In the next section we encounter the Cauchy sequence which has the advantage of being convergent without necessarily knowing what number the sequence
converges to. A key idea of this chapter will be the equivalence of these
two types of convergence.

Convergent   A sequence x_n converges, or has limit L, if for all $\epsilon > 0$ there
exists $N \in \mathbb{N}$ such that $n > N$ implies $|x_n - L| < \epsilon$. Then we write

$$\lim_{n \to \infty} x_n = L \quad \text{or} \quad \lim x_n = L \quad \text{or} \quad x_n \to L$$

We'll use the Archimedean Property to interpret statements like $N = 1/\epsilon$
to mean that N is next integer after the real number $1/\epsilon$.

Example 23 Prove that

$$\lim_{n \to \infty} \frac{1}{\sqrt{n}} = 0$$

Discussion The definition of convergence is

$$\forall \epsilon > 0 \; \exists N \in \mathbb{N} \text{ such that } n > N \implies |x_n - L| < \epsilon.$$

This says that for each epsilon we have to find an N. So essentially we solve the expression so that epsilon equals something with an n in it (and no epsilons). We start with what we want to show,

$$|x_n - L| < \epsilon$$

then put in the specifics for our problem

$$\left| \frac{1}{\sqrt{n}} - 0 \right| < \epsilon$$

then go about solving for n

$$\frac{1}{\sqrt{n}} < \epsilon$$

$$\frac{1}{\epsilon} < \sqrt{n}$$

$$\frac{1}{\epsilon^2} < n.$$

Now if $N > 1/\epsilon^2$ we should be good to go, provided the algebra works back the other way. We'll choose $N > 1/\epsilon^2$ and write the definition using this piece of information, and include the algebra steps to reach the desired conclusion. Often the proof is much shorter than the 'scratch' work.

Proof For all $\epsilon > 0$ let $N = 1/\epsilon^2$, then $n > N$ implies

$$n > 1/\epsilon^2 \implies \sqrt{n} > \frac{1}{\epsilon} \implies \frac{1}{\sqrt{n}} < \epsilon$$

and then we have our desired conclusion:

$$\left| \frac{1}{\sqrt{n}} - 0 \right| < \epsilon \; \square$$

Example 24   Prove that

$$\lim_{n\to\infty} \sqrt{n+1}-\sqrt{n}=0$$

Discussion   Here we'll attempt the same idea of solving for n in terms of ϵ but we'll encounter a difficulty along the way. First write the conclusion we need:

$$\left|\sqrt{n+1}-\sqrt{n}-0\right|<\epsilon$$

Note that $\sqrt{n+1}>\sqrt{n}>0$ so drop the absolute value and then we realize that squaring the left side doesn't get rid of the square root. A more creative approach is needed, found by trial and error (multiply by one):

$$\left(\sqrt{n+1}-\sqrt{n}\right)\cdot\left(\frac{\sqrt{n+1}+\sqrt{n}}{\sqrt{n+1}+\sqrt{n}}\right)<\epsilon,$$

$$\frac{1}{\sqrt{n+1}+\sqrt{n}}<\epsilon.$$

Then if we can find an expression we know is larger than the left hand side, by keeping it less than ϵ we'll have what we need. Notice that

$$\frac{1}{\sqrt{n+1}+\sqrt{n}}<\frac{1}{2\sqrt{n}}<\epsilon$$

works. Then we have that $1/(4\epsilon^2) < n$ and all we have to do is write the proof.

Proof   For all $\epsilon > 0$ let $N = 1/(4\epsilon^2)$, then $n > N$ implies $n > 1/(4\epsilon^2)$. hence $\epsilon^2 > \frac{1}{4n}$ and the following string of inequalities holds by algebra:

$$\left(\sqrt{n+1}-\sqrt{n}\right)\cdot\left(\frac{\sqrt{n+1}+\sqrt{n}}{\sqrt{n+1}+\sqrt{n}}\right)=\frac{1}{\sqrt{n+1}+\sqrt{n}}<\frac{1}{2\sqrt{n}}<\epsilon$$

$$\left|\sqrt{n+1}-\sqrt{n}-0\right|<\epsilon.$$

Therefore

$$\lim_{n\to\infty} \sqrt{n+1}-\sqrt{n}=0\;_\square$$

Limits are Unique   Suppose $x_n \to a$ and $x_n \to b$, then $a = b$.

Discussion   It might seem odd to need to prove this but until we do, who knows if it is true. If we had defined convergence so that

$$a_n = 1, 0, 1, 0, 1, 0, 1, 0, \ldots$$

was convergent, then it would reasonably converge to two different numbers. This will be our starting point for the proof. For the conclusion we'll need to appropriately interpret the statement $|b - a| < \epsilon$. The definition of convergence starts with, for all $\epsilon > 0$, so when we encounter $|b - a| < \epsilon$ our conclusion is that there is no distance between a and b. Hence they are the same.

Proof   Since $x_n \to a$ and $x_n \to b$ we have that for all $\epsilon > 0$ there exists $N \in \mathbb{N}$ such that $n > N$ implies both

$$|x_n - a| < \epsilon \text{ and } |x_n - b| < \epsilon.$$

We can add these two inequalities to get

$$|x_n - a| + |b - x_n| < 2\epsilon,$$

and then apply the triangle inequality

$$|b - a| = |x_n - a + b - x_n| \leq |x_n - a| + |b - x_n| < 2\epsilon.$$

So that $|b - a| < 2\epsilon$ and hence $a = b$. Therefore limits are unique $_\square$

Comment   Notice that at the end we have 2ϵ instead of ϵ. This doesn't really matter because the epsilon is arbitrarily small. The statement starts with all of them, so if we get a multiple of one it still works. To see this, in most proofs I'll start with ϵ over something, and in this proof if you start with $\epsilon/2$ you'll get a result with just ϵ. Also, even though there are two sequences a_n and b_n, we've taken a shortcut by picking one N for both of them, when there is probably an N_1 with a_n and an N_2 with b_n. This is spelled out in the next proof.

The next two theorems are presented separately, but together they are called linearity which means that for two convergent sequences $a_n \to a$ and $b_n \to b$, we have $\lim_{n \to \infty} k(a_n + b_n) = ka + kb$.

Sum of Sequences   Let $a_n \to a$ and $b_n \to b$. Then $(a_n + b_n) \to (a + b)$.

Proof   Since a_n is convergent, for all $\epsilon > 0$ there exists $N_1 \in \mathbb{N}$ such that $n > N_1$ implies
$$|a_n - a| < \epsilon/2.$$
Similarly, there is a N_2 for b_n with
$$|b_n - b| < \epsilon/2.$$
Let $N = \max\{N_1, N_2\}$ and then add the inequalities to get that for all $\epsilon > 0$ there is an $N \in \mathbb{N}$ such that $n > N$ implies
$$|a_n - a| + |b_n - b| < \epsilon.$$
Next apply the triangle inequality to get
$$|a_n - a + b_n - b| \le |a_n - a| + |b_n - b| < \epsilon,$$
and rearranging terms yields
$$|a_n + b_n - (a + b)| \le |a_n - a| + |b_n - b| < \epsilon.$$
Therefore $(a_n + b_n) \to (a + b)$ $\square$

Constant Times a Sequence   Let $a_n \to a$ and $k \in \mathbb{R}$. Then $ka_n \to ka$.

Proof   Since a_n is convergent, for all $\frac{\epsilon}{|k|} > 0$ there exists $N \in \mathbb{N}$ such that $n > N$ implies
$$|a_n - a| < \frac{\epsilon}{|k|}.$$
Multiply by $|k|$ to get
$$|k||a_n - a| < |k|\frac{\epsilon}{|k|},$$
$$|ka_n - ka| < \epsilon.$$
Therefore $ka_n \to ka$ $\square$

The next idea is that if a sequence is between two convergent sequences, converging to the same point, then all three sequences converge to the same point.

Squeeze lemma Consider three sequences such that $a_n \leq s_n \leq b_n$ and that $\lim a_n = \lim b_n = s$. Prove $\lim s_n = s$.

Proof From the given limits we have that for each $\epsilon > 0$ there exists N_1 so that
$$n > N_1 \implies |a_n - s| < \epsilon$$
and that there exists N_2 so that
$$n > N_2 \implies |b_n - s| < \epsilon.$$

Let $N = \max\{N_1, N_2\}$. Then for $n > N$ we have both the inequalities below:
$$|s - a_n| < \epsilon \quad \text{and} \quad |b_n - s| < \epsilon$$
which we can manipulate to yield
$$-\epsilon < a_n - s < \epsilon \quad \text{and} \quad -\epsilon < b_n - s < \epsilon,$$
$$s - \epsilon < a_n \quad \text{and} \quad b_n < \epsilon + s,$$
and then apply the given inequality to combine them as follows:
$$s - \epsilon < a_n \leq s_n \leq b_n < \epsilon + s.$$

Since we want to say that s_n converges, we'll drop the a_n and b_n to yield
$$s - \epsilon < s_n < \epsilon + s,$$
$$-\epsilon < s_n - s < \epsilon,$$
and finally
$$|s_n - s| < \epsilon.$$
Since this is true for all $n > N$ we have that $\lim s_n = s$ □

Bounded   A sequence x_n is bounded if there exists $M \in \mathbb{R}$ such that for all n, $|x_n| \leq M$.

In certain cases one might only need bounded above, $x_n \leq M$, or only bounded below, $x_n \geq M$. It turns out (hopefully as no surprise) that convergent sequences are bounded.

Convergent $\Longrightarrow$ Bounded   Let x_n be a convergent sequence, that is

$$\lim_{n \to \infty} x_n = L,$$

then x_n is bounded.

Discussion   So we have the statement of convergence:

$$\forall \epsilon > 0 \ \exists N \in \mathbb{N} \ \text{ such that } \ n > N \implies |x_n - L| < \epsilon,$$

and we need to show bounded

$$\exists M \in \mathbb{R} \ \text{ such that } \ \forall n \ |x_n| < M$$

So we need to take the convergence as given and define our choice of M in terms of the given letters of convergence. Convergence says any ϵ will do, so we'll pick a convenient one, $\epsilon = 1$, then

$$|x_n - L| < 1$$

and we'll get the L out of the absolute value with some algebra. Recall that this statement is only true for $n > N$ and so we have a finite number of cases where x_n may be farther away. We'll account for them using a maximum function of their absolute value. Note that L could be negative as well.

Proof   Let $\epsilon = 1$ and we have that $|x_n - L| < 1$ which means that $n > N$ implies

$$-1 < x_n - L < 1,$$

$$-1 + L < x_n < L + 1.$$

Let $M = \max\{|L| + 1, |x_1|, |x_2|, \ldots, |x_N|\}$. Then $|x_n| \leq M$ which means x_n is bounded $_\square$

Product of Sequences If $a_n \to a$ and $b_n \to b$ then $a_n b_n \to a b$.

Discussion In the proof of the squeeze lemma and additivity of sequences we were able to add inequalities or use transitivity. For this proof we'll have to figure out how to get from $|a_n - a|$ and $|b_n - b|$ to $|a_n b_n - a b|$. By trial and error from $|a_n b_n - a b|$ we can find a creative way to add zero that works, $0 = a_n b - a_n b$. Then

$$a_n b_n - a b = a_n b_n - a_n b + a_n b - a b = a_n(b_n - b) + b(a_n - a).$$

Since a_n is convergent, it is bounded. So the right side is the sum of two quantities, each of them a bounded quantity times an arbitrarily small one, hence they can be made arbitrarily small.

Proof From the given limits we have that for each $\epsilon > 0$ there exists N_1 so that

$$n > N_1 \implies |a_n - s| < \epsilon$$

and that there exists N_2 so that

$$n > N_2 \implies |b_n - s| < \epsilon.$$

Let $N = \max\{N_1, N_2\}$. Note that

$$a_n b_n - a b = a_n(b_n - b) + b(a_n - a),$$

and so by the absolute value properties and triangle inequality

$$|a_n b_n - a b| = |a_n(b_n - b) + b(a_n - a)| \le |a_n||b_n - b| + |b||a_n - a|.$$

Since a_n is convergent, it is bounded, say by M. Then using ϵ from the given convergence we have

$$|a_n b_n - a b| < M\epsilon + |b|\epsilon = \epsilon(M + |b|).$$

Therefore $a_n b_n \to a b$. Note that if we had started with $\frac{\epsilon}{M+|b|}$ we would have an ϵ alone at the conclusion $\square$

Next we introduce the idea of monotonic, which we can combine with boundedness to say a sequence is convergent.

Monotonic   A sequence is monotonic if for all n it is increasing $x_n \leq x_{n+1}$ (but not necessarily strictly increasing), or decreasing $x_{n+1} \leq x_n$ (but not necessarily strictly decreasing).

Bounded + Monotonic $\Longrightarrow$ Convergent   Let x_n be a bounded and monotonic sequence. Then x_n is convergent.

Discussion   Since x_n is bounded we have an M such that

$$|x_n| \leq M.$$

And since its monotonic we'll do one case, $x_n \leq x_{n+1}$. We need to show

$$\forall \epsilon > 0 \ \exists N \in \mathbb{N} \ \text{ such that } \ n > N \Longrightarrow |x_n - L| < \epsilon.$$

We can write the boundedness as

$$-M \leq x_n \leq M,$$

but to say the sequence converges, we need the number that it converges to. The sequence $y_n = -1, -1/2, -1/3, -1/4, \ldots$ is bounded above by 57 but 57 has nothing to do with its limit. We need the least upper bound, but this is defined for a set of numbers, not a sequence. Luckily we're free to construct a set from a sequence. So let

$$X = \{x_n | x_n \text{ is bounded and monotonic}\}.$$

Then since the sequence x_n is bounded the set X is bounded as well. Since it is a bounded set of real numbers it has a supremum,

$$a = \sup X.$$

Now our guess is that x_n converges to a. We'll use the meaning of upper bound to conclude that for any $\epsilon > 0$

$$x_n \leq a < a + \epsilon.$$

Then use the meaning of least upper bound to conclude that for some N we have that $n > N$ implies

$$a - \epsilon < x_n.$$

Putting these together yields the desired conclusion, that is

$$a - \epsilon < x_n < a + \epsilon \Longrightarrow |x_n - a| < \epsilon.$$

Bounded + Monotonic $\implies$ Convergent Let x_n be a bounded and monotonic sequence. Then x_n is convergent.

Proof Let

$$X = \{x_n \mid x_n \text{ is bounded and monotonic}\}.$$

Then since x_n is bounded, X is bounded (it only contains elements from the bounded sequence). Since X is a bounded set of real numbers it has a supremum that is a real number,

$$a = \sup X.$$

Let ϵ be a real number such that $\epsilon > 0$, then since a is an upper bound we have that

$$x_n \le a < a + \epsilon.$$

Since a is the least upper bound, a slightly lower number, $a - \epsilon$ will be less than some particular x_N,

$$a - \epsilon < x_N.$$

And since x_n is monotonic, any other term later in the sequence is greater, that is

$$a - \epsilon < x_N \le x_n.$$

Combining inequalities we have that for all $n > N$,

$$a - \epsilon < x_N \le x_n \le a < a + \epsilon,$$

$$-\epsilon < x_n - a < \epsilon,$$

$$|x_n - a| < \epsilon.$$

Therefore

$$\lim_{n \to \infty} x_n = a.$$

There remains another case. In this proof we only considered the monotonically increasing case $_\square$

The monotonically decreasing case requires the existence of the infimum of a set. Note that we have not yet established the existence of the infimum as a real number yet. The completeness axiom was only for the supremum. Since the decreasing case follows the same structure of proof, we'll leave it to the reader. However, we will prove the existence of the infimum next.

Corollary to Completeness Axiom   Every non-empty set, S, of real numbers, bounded below, has a greatest lower bound, called the infimum of S, or $\inf S$, and $\inf S \in \mathbb{R}$.

Discussion   Consider the following set, S, on the real line

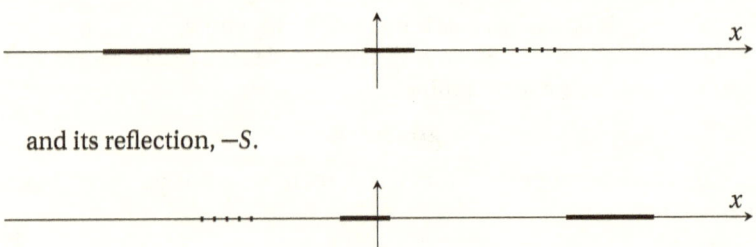

and its reflection, $-S$.

It appears that

$$\inf S = -\sup{-S}.$$

Take a moment to relate those math symbols to the above pictures. For this to be true we must first establish that $-\sup{-S}$ is a lower bound, then establish that it is the greatest lower bound. In order to do this we use the notation

$$-S = \{-s \mid s \in S\}$$

and start with $s \in S$ to get that $-\sup{-S}$ is a lower bound. For the greatest lower bound, we'll let t be any lower bound and then use inequality rules with the definition of supremum to show that $-\sup{-S}$ is greater than or equal to any other lower bound. By the completeness axiom the supremum (even of a negative set) exists as a real number and then by the additive inverse property $-\sup{-S}$ is a real number as well.

Corollary to Completeness Axiom   Every non-empty set, S, of real numbers, bounded below, has a greatest lower bound, called the infimum of S, or $\inf S$, and $\inf S \in \mathbb{R}$.

Proof   Let $S \subset \mathbb{R}$ be bounded below. Then there exists $m \in \mathbb{R}$ such that $m \le s$ for all $s \in S$. But then

$$-m \ge -s \text{ for all } s \in S.$$

And since $-S$ is defined as $-S = \{-s \mid s \in S\}$, we conclude that $-S$ is bounded above by $-m$. Let $-s \in -S$ and then by definition of supremum we have that

$$-s \le \sup -S,$$

$$s \ge -\sup -S,$$

and $-\sup -S$ is a lower bound of S. Now let t be any lower bound of S. Then

$$t \le s \ \forall \ s \in S,$$

$$-t \ge -s \ \forall \ s \in S,$$

and since $s \in S$ means $-s \in -S$ we have that

$$-t \ge -s \ \forall -s \in -S$$

which means that $-t$ is an upper bound of $-S$. And therefore

$$-t \ge \sup -S,$$

and since $\sup -S$ is a real number,

$$t \le -\sup -S.$$

Since t is any lower bound and $-\sup -S$ is greater, it is the greatest lower bound. Hence

$$\inf S = -\sup -S.$$

Finally, since $\sup -S$ is a real number, its additive inverse, $-\sup -S$ is real as well $\square$

2.2   Cauchy Sequences and Subsequences

Our last big idea was that a bounded monotonic sequence converges. Notice that the bounded and monotonic property don't include the number to which the sequence converges. But there are sequences that are bounded and not monotonic that converge, like

$$x_n = \frac{1}{1}, \frac{-1}{2}, \frac{1}{3}, \frac{-1}{4}, \frac{1}{5}, \dots$$

So we next define the Cauchy sequence (whose definition does not include the limit) because it turns out that the Cauchy sequences are convergent, yet they are of a more fundamental nature. The idea is that if the difference between each successive pair of terms gets smaller and smaller, then the sequence converges.

Cauchy   A sequence is Cauchy if for all $\epsilon > 0$ there exists $N \in \mathbb{N}$ such that $n, m > N$ implies

$$|x_m - x_n| < \epsilon$$

Convergent $\implies$ Cauchy   Any convergent sequence of real numbers is Cauchy.

Proof   Let $x_n \to x$. Then to get the subscript m we can say the same thing using m, that is $x_m \to x$ (note that m and n are just different indices for the same sequence). Now for all $\epsilon > 0$ there exists $N \in \mathbb{N}$ such that $n, m > N$ implies both

$$|x_n - x| < \epsilon/2 \ \text{ and } \ |x_m - x| < \epsilon/2.$$

Now add inequalities and apply the triangle inequality:

$$|x_n - x_m| = |x - x_n + x_m - x| \le |x - x_n| + |x_m - x| < \frac{\epsilon}{2} + \frac{\epsilon}{2} = \epsilon.$$

Therefore $|x_n - x_m| < \epsilon$ and the sequence is Cauchy $\square$

The proof in the other direction is not as straightforward and we'll need several other ideas that we'll develop first.

Cauchy $\implies$ Bounded   Every Cauchy sequence is bounded.

Discussion   We'll use the proof that every convergent sequence is bounded as a model. There we let $\epsilon = 1$ and then replaced the absolute value with some inequalities. We do the same here, however we'll need to figure out what the bound should be. In our model the bound was the maximum of $|L| + 1$ and the absolute value of all the terms up to $|x_N|$. We do the same using x_{N+1} in place of L.

Proof   Let x_n be a Cauchy sequence. Then for all $\epsilon > 0$ there exists $N \in \mathbb{N}$ such that $n, m > N$ implies

$$|x_n - x_m| < \epsilon.$$

Let $\epsilon = 1$ and then when $m = N + 1$ we have

$$|x_n - x_{N+1}| < 1,$$

$$-1 < x_n - x_{N+1} < 1,$$

$$-1 + x_{N+1} < x_n < 1 + x_{N+1}.$$

So that for all $n > N$, x_n is bounded by $x_{N+1} \pm 1$. Since $N + 1$ is finite the entire sequence is bounded by

$$M = \max\{|x_{N+1}| + 1, |x_0|, |x_1|, \ldots, |x_N|\} \; \square$$

The definition of Cauchy sequence has in it the seed for our next idea, subsequences. The use of the notation x_n and x_m indicates that probably $m \neq n$. However this isn't required anywhere. Are these two different sequences, or are they different terms in the same sequence? We should think of them as different terms in the same sequence, and then we can use the notation x_{n_k} if we want to talk about a sequence that gets all its terms (in order) from another sequence. We make this formal on the next page.

Subsequence   Let n_k be a strictly increasing sequence of natural numbers,

$$n_1 < n_2 < \cdots$$

Then from any sequence x_n we have a subsequence x_{n_k}.

The next theorem is a little bit stronger than what we will need when we get to proving that Cauchy sequences are convergent. However, its proof requires a bit of creativity to come up with.

Sequence $\Longrightarrow$ Monotonic Subsequence   Every sequence has a monotonic subsequence.

Discussion   If our sequence is monotonic, there isn't much to do, so we should think about the non-monotonic case where there terms alternate back and forth, perhaps like:

$$1, 0, -1, 1, 0, -1, 1, 0, -1 \ldots$$

But then then if we just take the terms on one side (of zero in this case) then we have a monotonic subsequence. But this is just one case. Our sequence could be anything. The needed observation is about the number of 'largest terms', that is, a term that bounds all the remaining terms of the sequence. There could be a finite number of these terms, or infinitely many, and thus there are two cases to consider. If there are infinitely many of these 'largest terms' then they make a monotonic subsequence. If there are finitely many of them we can start with the last one, and we know there is not another 'largest term'. We use this to find the next term in our subsequence. First we need to rename this 'largest term' to something more appropriate:

Dominant Term   Let x_n be a sequence. A particular number, x_m, in the sequence is a dominant term if

$$x_n < x_m \text{ for all } n > m$$

Sequence $\Longrightarrow$ Monotonic Subsequence   Every sequence has a monotonic subsequence.

Proof   Let x_n be a sequence. Then there are infinitely many dominant terms, or finitely many:

C1 Infinitely many dominant terms: Let x_{m_k} be the kth dominant term. Then x_{m_k} is a subsequence and from the dominant property we have

$$x_{m_1} < x_{m_2} < x_{m_3} < \cdots$$

so that it is monotonic as well.

C2 Finitely many dominant terms: Let x_N be the last dominant term. Then for all $n > N$ the terms are not dominant. So there exists x_{n_1} such that

$$x_{n_1} \geq x_{N+1}.$$

Since x_{n_1} is not dominant there exists another term, x_{n_2} such that

$$x_{n_2} \geq x_{n_1}.$$

Since x_{n_2} is not dominant there exists another term, x_{n_3} such that

$$x_{n_3} \geq x_{n_2}.$$

In this way we construct the subsequence so that

$$\cdots \geq x_{n_5} \geq x_{n_4} \geq x_{n_3} \geq x_{n_2} \geq x_{n_1}.$$

From the above line of inequalities we see that the subsequence is monotonic.

Hence, in either case a sequence has a monotonic subsequence. It is left to the reader to consider the case with no dominant terms $\square$

Bolzano-Weirerstrass   Every bounded sequence has a convergent subsequence.

Proof   Let x_n be a bounded sequence. By the previous theorem it has a monotonic subsequence, x_{n_k}. Since x_{n_k} comes from a bounded sequence it is bounded as well. Since x_{n_k} is bounded and monotonic, it is convergent $\square$

Cauchy $\Longrightarrow$ Convergent   Let x_n be a Cauchy sequence, then x_n is convergent.

Discussion   With the related results on sequences so far we can begin to put them together in several ways, but not quite to get the conclusion we need. The missing piece is that if we have a convergent subsequence of a Cauchy sequence, then the Cauchy sequence is convergent. We can get this by applying their definitions with the triangle inequality.

Proof   Let x_n be a Cauchy sequence. Then x_n is bounded and by Bolzano-Weirerstrass it has a convergent subsequence, x_{n_k}. Since x_n is Cauchy we have that for all $\epsilon > 0$ there exists $N_1 \in \mathbb{N}$ such that $n, m > N_1$ implies

$$|x_n - x_m| < \frac{\epsilon}{2}.$$

Also since the subsequence is convergent, for all $\epsilon > 0$ there exists $N_2 \in \mathbb{N}$ such that $n_k > N_2$ implies

$$|x_{n_k} - x| < \frac{\epsilon}{2}.$$

Let $p = \max\{m, n_k\}$ and then add inequalities to get

$$|x_p - x| + |x_n - x_p| < \epsilon,$$

and then apply the triangle inequality.

$$|x_n - x| = |x_p - x + x_n - x_p| \leq |x_p - x| + |x_n - x_p| < \epsilon.$$

Hence for $n > \max\{N_1, N_2\}$ we have that

$$|x_n - x| < \epsilon$$

and x_n is convergent $\square$

2.3 Real Numbers Defined

> Quite frankly, no mature mathematician thinks of an equiv-
> alence class of Cauchy sequences of rationals every time the
> word 'real number' appears.
> –Robert Strichartz in The Way of Analysis

This section is here not for any mathematical content, but to give you the words to use if someone happens to ask you what a real number is. For the words to make sense we need to define them, so lets get on with it. A relation is something like the equals sign, or one of the inequality signs, or some symbol that says two things are related. For the generic case of any relation we'll use the symbol $\mathscr{R}$. Now for the most important kind, equivalence relation.

Equivalence Relation   An equivalence relation is a relation, $\mathscr{R}$ on elements of a set, S with the following properties for any elements $a, b, c \in S$.

- It is *reflexive*: $a\mathscr{R}a$.

- It is *symmetric*: $a\mathscr{R}b \implies b\mathscr{R}a$.

- It is *transitive*: $a\mathscr{R}b$ and $b\mathscr{R}c \implies a\mathscr{R}c$.

With our real numbers notice that since $a < a$ is false, strict inequality is not an equivalence relation (violates reflexive property). Also notice that $a \le b \nimplies b \le a$ so that non-strict inequality is not an equivalence relation either (violates symmetric property).

Partition of a Set   A partition is a collection of subsets, say X_n of X such that each element from X is in exactly one subset. These subsets are called *classes*. The notation for the class will be any element of the class with a bar over its symbol. So if c is an element of the class, $\bar{c}$ is the whole class.

For example, the positive integers can be partitioned into the even and odd classes. It turns out that an equivalence relation induces a partition on a set. For the even and odd class just mentioned the equivalence relation is divisibility by two–the remainder is zero or one. It is natural to use one of the elements from the class in order to identify it. So in this example we would have the two classes:

$$\bar{0} = \{0, 2, 4, 6, \ldots\}$$

$$\bar{1} = \{1, 3, 5, 7, \ldots\}$$

Since this partition of $\mathbb{N}$ is induced by the equivalence relation, the classes are called *equivalence classes*.

Equivalence Relation $\implies$ Partition   Let $\mathcal{R}$ be an equivalence relation on a set S. Then $\mathcal{R}$ induces a partition of S.

Discussion   This proof is of a different nature than any of the other proofs in the book because it comes from algebra and set theory. What it is doing is using the notation of a cell and the properties of an equivalence relation to conclude that S is partitioned. Then a similar argument works in the other direction.

Proof   Let $\bar{a}$ be a cell of S. Then if $a \in S$ we have that $a \in \bar{a}$ so that each element of S is in a cell. To see this suppose $b \in S$ then either $a \mathcal{R} b$ or $b \in \bar{b}$. If $b \in \bar{b}$ then it is in a cell and we are done. If not then we need to show $\bar{a} = \bar{b}$. This is equality of sets which is shown by containment.

Let $x \in \bar{a}$ then since the relation defines the cell, $x \mathcal{R} a$ and from above $a \mathcal{R} b$. Now by transitivity $x \mathcal{R} b$ so that $x \in \bar{b}$ and we have $\bar{a} \subset \bar{b}$. For the other direction let $y \in \bar{b}$ and then $y \mathcal{R} b$. Also, from above $a \mathcal{R} b$ so that $b \mathcal{R} a$, and by transitivity we have $y \mathcal{R} a$. Therefore $y \in \bar{b}$ and $\bar{b} \subset \bar{a}$. Hence $\bar{a} = \bar{b}$.

We have shown that each element of S is in a cell, hence S is partitioned $\square$

Partition $\Longrightarrow$ Equivalence Relation   A partition of S defines an equivalence relation $\mathscr{R}$.

Proof   Suppose S is partitioned into cells. Define $x \mathscr{R} y$ if they are in the same cell. Then we have $x \mathscr{R} x$ because x is in the cell it is in. Hence $\mathscr{R}$ is reflexive. For symmetric note that $x \mathscr{R} y$ means x and y are in the same cell so that $y \mathscr{R} x$. For transitive note that $x \mathscr{R} y$ and $y \mathscr{R} z$ mean that x and y are in the same cell, and that y and z are in the same cell, hence x and z are in the same cell and therefore $x \mathscr{R} z$. We conclude that a partition induces an equivalence relation $\square$

Real Number   A real number is the equivalence class of all Cauchy sequences of rational numbers converging to the same point.

Consider these three sequences of rational numbers that all converge to $\sqrt{2}$:

$$\sqrt{2} = \begin{cases} x_n = & 1, \frac{3}{2}, \frac{11}{8}, \frac{69}{48}, \frac{261}{192}, \ldots \\ y_n = & 1, \frac{3}{2}, \frac{17}{12}, \frac{577}{408}, \ldots \\ z_n = & 3, \frac{11}{6}, \frac{193}{132}, \frac{72079}{50952}, \ldots \\ \vdots & \end{cases}$$

They are in the same equivalence class of Cauchy sequences of rational numbers. They are three equivalent ways to represent the real number $\sqrt{2}$. Every other real number has its class of Cauchy sequences of rational numbers as well. There are uncountably many of them.

There is another way to describe the real numbers developed by Richard Dedekind, called Dedekind cuts. Note that both the ideas of Dedekind cuts and equivalence classes of Cauchy sequences are consistent with the real numbers from the axiomatic approach. This illustrates how the mathematical properties of objects are important and useful regardless of how one might construct the objects. On the other hand, the process of constructing objects is good practice to develop the skills that are useful in manipulating the objects in accordance with their properties. We'll continue that practice with the study of continuity next.

Chapter 3

Continuity

Continuity is a tool to say that a function is smooth (perhaps unbroken would be better at first, but doesn't work later on). There are several types of continuity, but we only need the first two in order to define the integral so that is as far as we go. They are called continuous, and uniformly continuous.

A decent calculus teacher probably told you that what makes a function continuous is that if two numbers are close together in the domain of the function, then their image, or the two points they map to, are close to each other in the range of the function. The opposite turns out to be the way continuity is actually defined.

The definition that always applies is that for a function to be continuous, the pre-image of an open set must be open. This means that for any open set, U, in the range, there must be an open set in the domain that the function maps to U. Think of the open set as containing points that are close to each other. For our purposes the set is a a subset of the real number line, and we have some experience with sequences, so we'll initially define closed sets in terms of sequences.

3.1 Closed and Open Sets

By now we should be getting comfortable with the way sequences work and some of their properties. In this section we'll use sequences to describe closed sets which on the real line take the form of intervals like $[a, b]$, as well as intersections and unions of closed intervals. However, our experience does give some intuition about why the words open and closed are used. Consider some x in the open interval $(1, 2)$. No matter which x you chose, the door is open to find another point on either side of x that is still in the interval $(1, 2)$. However, if we tried this with the closed interval $[1, 2]$ and you happen to choose 1 or 2 for your point, you are closed off on one side from picking another point in the interval. Hence it is closed. Now for a definition of a closed set of real numbers.

Closed Set of Real Numbers Let $X \subset \mathbb{R}$. Then X is closed if every convergent sequence of numbers from X converges to a point in X.

Union and Intersection of Closed Sets We'll use the fact that intersections and finite unions of closed sets are closed without proof.

Example 25 Consider the following sequence from earlier

$$x_n = 1, 0, -1, 1, 0, -1, 1, 0, -1, \ldots$$

Note that we can think of this sequence as coming from the set

$$\{1, 0, -1\}.$$

Even though x_n is not convergent, the sequence

$$y_n = 1, 1, 1, 1, 1, 1, 1, \ldots$$

is a convergent sequence from the set $\{1\}$. Similar observations are made for sequences from $\{0\}$, and $\{2\}$ so that they are each closed sets. Then by the union of closed sets, their union is closed, hence $\{-1, 0, 1\}$ is a closed set.

Discussion   There is something about this sequence x_n from above that we don't quite have the vocabulary to describe yet. It seems as if it should be convergent to three points simultaneously, but the definition of convergence doesn't allow that. It even has an infinite number of terms close to 1 and to 0 and to -1. Surely there is a name for this. Indeed, it is called a limit point of a sequence.

Limit Point of a Sequence   Let x_n be a sequence. Then if for all $\epsilon > 0$ there exists b such that there are an infinite number of terms, x_j with

$$|x_j - b| < \epsilon,$$

then b is a limit point of x_n

There may be multiple limit points for a sequence as there are in the sequence $x_n = 1, 0, -1, 1, 0, -1 \ldots$ on the previous page.

Example 26   Consider the following sequence

$$x_n = \frac{3}{2}, \frac{1}{2}, \frac{7}{4}, \frac{1}{4}, \frac{9}{8}, \frac{1}{8}, \frac{17}{16}, \frac{1}{16}, \ldots$$

It has two limit points, one and zero. What are the subsequences associated with each limit point?

limsup and liminf   Next we'll present the tool to pluck out the largest and smallest limit points in a sequence. It works like this. Take all the terms of a sequence and put them in a set. Recall that a set lists each element only once, while a sequence could be one number repeated infinitely many times.

$$\left\{ 1, 3, \frac{3}{2}, \frac{1}{2}, \frac{5}{4}, 4, \frac{1}{4}, \frac{9}{8}, \frac{1}{8}, \frac{17}{16}, \frac{1}{16}, \ldots, \frac{2^n + 1}{2^n}, \frac{1}{2^n} \ldots \right\}.$$

Since the set is bounded, and not empty, the supremum exists. We'll start with that

$$\sup \left\{ 1, 3, \frac{3}{2}, \frac{1}{2}, \frac{5}{4}, 4, \frac{1}{4}, \frac{9}{8}, \frac{1}{8}, \frac{17}{16}, \frac{1}{16}, \ldots, \frac{2^n + 1}{2^n}, \frac{1}{2^n} \ldots \right\} = 4.$$

But 4 has nothing to do with the limit point here. We want to be farther out in the sequence, so consider just those $n > N$. For $N = 4$.

$$\sup_{n>N}\left\{\frac{17}{16},\frac{1}{16},\ldots,\frac{2^n+1}{2^n},\frac{1}{2^n}\ldots\right\}=\frac{17}{16}.$$

Now if we consider the limit as N increases without bound we have

$$\lim_{N\to\infty}\sup_{n>N}\left\{1,3,\frac{3}{2},\frac{1}{2},\frac{5}{4},4,\frac{1}{4},\frac{9}{8},\frac{1}{8},\frac{17}{16},\frac{1}{16},\ldots,\frac{2^n+1}{2^n},\frac{1}{2^n}\ldots\right\}=1.$$

The shorter version is

$$\limsup\left\{1,3,\frac{3}{2},\frac{1}{2},\frac{5}{4},4,\frac{2^n+1}{2^n},\frac{1}{2^n}\ldots\right\}=1.$$

And the smallest limit point of a sequence comes from the $\liminf$ in a similar manner so that

$$\liminf\left\{1,3,\frac{3}{2},\frac{1}{2},\frac{5}{4},4\frac{2^n+1}{2^n},\frac{1}{2^n}\ldots\right\}=0.$$

Limit points are related to subsequences in a direct way, but before we detail that relationship we need to make the notion of infinity more precise. The symbol, ∞ is what we write when a quantity increases without bound. Similarly we write $-\infty$ when a quantity decreases without bound. Neither of these symbols are numbers in $\mathbb{R}$ so no theorem about real numbers can be applied to them. However there is the extended real line, $\mathbb{R}\cup\{-\infty,\infty\}$ with the two additional points.

Another way to describe the real numbers is to say that the real numbers constitute the limit points of every cauchy sequence of rational numbers. The precise name for this concept in general is completion.

Complete Given a set A, and a metric, $d(x,y) : A \times A \to \mathbb{R}$; if every cauchy sequence in A is convergent to a point in A, then A is complete.

This definition is given here because it is related to topology, but we do not refer to it until chapter 6 where we discuss sequences of functions. The idea turns out to be quite powerful because it ties the real numbers (the completion of the rational numbers) to any other set with a defined metric. In this way functions can be thought of as points in a set. Then we are able to make statements about the existence of the functions if they come from a complete set or not. If they exist, then it is worth looking for them. If they don't exist, then don't bother to look for them. Back to limit points...

Limit Point $\Longleftrightarrow$ Convergent Subsequence   Let x_n be a sequence. Then b is a limit point of x_n if and only if there is a subsequence x_{n_k} of x_n converging to b.

Discussion   Once we write the definitions of convergence and limit point we observe that if $j = n_k$ were are in good shape. We also need to observe that both definitions are about an infinite number of terms in the sequence.

Proof   Let x_{n_k} be a subsequence converging to a point b. Then for all $\epsilon > 0$ there exists $N \in \mathbb{N}$ such that $n_k > N$ implies

$$|x_{n_k} - b| < \epsilon.$$

Then with $j = n_k$ in the definition of a limit point, we have that b is a limit point of a sequence. Now for the other direction, let b be a limit point of x_n. Then there are an infinite number of terms, x_j, with

$$|x_j - b| < \epsilon$$

Let x_{n_1} be the first such term, and x_{n_2} the second and x_{n_3} the third and so on. Then we have the subsequence x_{n_k} with

$$|x_{n_k} - b| < \epsilon$$

so that $x_{n_k} \to b$ as required $\square$

Example 27   To see why only finite unions of closed sets are closed, consider the union

$$\bigcup_{n \in \mathbb{N}} [0, 1 - \frac{1}{n}] = [0, 1).$$

So an infinite union of closed sets is not necessarily closed, or even open.

Example 28 On the other hand an example of an infinite intersection is as follows:

$$\bigcap_{n\in\mathbb{N}} [0, \frac{1}{n}] = \{0\}$$

and results in a closed set as required.

Now that we have described the closed sets, we'll move on to the open sets. Note that the presentation here is only for sets of real numbers, and we are using the standard topology. One could study topology, and select different characteristics to define closed and open sets. Those characteristics would be described in the basis of the topology. The standard basis used on real numbers is given below as our starting point for open sets.

Basis The basis of the standard topology on the real line is the collection of open intervals

$$\mathscr{B} = \{(a,b) \subset \mathbb{R}| \ a < b\}$$

Open Set An open set is any union of basis elements.

This next property called containment is the most important thing about open sets, and the most fundamental idea of topology on the real line.

Containment A non-empty open set of real numbers, U has the following property: Let $x \in U$, then there is a basis element such that $x \in (a,b) \subset U$. Furthermore, let $c = \frac{x+a}{2}$ and $d = \frac{x+b}{2}$. Then $x \in (c,d) \subset (a,b)$.

Example 29 $\emptyset$ is open because it is in every set, and it is closed because every sequence from it converges (there are none to check). $\mathbb{R}$ is open because it is the union of every basis element. Also $\mathbb{R}$ is closed because it contains the limit of every convergent sequence from $\mathbb{R}$.

Example 30 The interval $(1,2]$ is neither open nor closed. It is not closed because the following convergent sequence is in the interval but its limit, 1 is not:

$$x_n = 2, \frac{3}{2}, \frac{7}{4}, \frac{9}{8}, \frac{17}{16}, \dots$$

It is not open because there is no basis element containing 2 that is contained in (1,2].

In the next theorem we merge the ideas of topology and the density of the rational numbers in the real numbers. This gives us a strong tool to use later on.

Open Sets Contain Rationals   Every non empty open set of real numbers contains a rational number.

Proof   Let U be an open set. Then there is a basis element (a,b) contained in U. By density of the rationals there exists $q \in \mathbb{Q}$ such that $a < q < b$. Hence $q \in (a,b) \subset U$. Since U is any open set, every nonempty open set of real numbers contains a rational number $_\square$

Discussion   There is a subtle interpretation of the above statement that we present as a corollary. It is simply the contrapositive of the above theorem, but its interpretation is that no set of irrational numbers is open.

Corollary   A set of real numbers without any rational numbers is empty or not open.

In the next chapter we start Continuity and then encounter some difficulties that require us to use these ideas of open and closed sets. We'll end this section with two examples:

$$\left\{ 0, \frac{1}{n} \mid n \in \mathbb{N} \right\} , \left\{ \frac{1}{n} \mid n \in \mathbb{N} \right\}$$

Both of these sets are not open because they don't contain any basis elements. One is closed because it contains all of its limit points, while the other is not closed since it does not contain all of its limit points.

This section has presented a few principles without proof. This is regrettable, but necessary without an extensive chapter on topology. An introductory course in topology proves all the principles stated here from the four axioms of a topological space (or the principle is an axiom itself).

This page intentionally left blank.

3.2 Continuous Functions

When we think of a continuous function as associating points that are
close in the range to points that are close in the domain, and do this in
the framework of sequences of numbers, a few observations can help:

- a convergent sequence gives us lots of numbers that are close for
  large n. Think of them in the range of a function.

- The numbers get to be close to the point of convergence.

- Then if we look for where the sequence in the range comes from, it
  should always come from a convergent sequence in the domain.

These ideas motivate the following definition:

Sequential Continuity at a Point   Let $f : D \rightarrow R$ be a real valued function of real numbers and x_n any sequence converging to $x_0 \in D$. Then if
$f(x_n) \rightarrow f(x_0)$ for *every* sequence converging to x_0, f is continuous at x_0.

Example 31   The function

$$f(x) = \begin{cases} 1 & x < 0 \\ x & x \geq 0 \end{cases}$$

is not continuous. Consider $x_n = -1, -1/2, -1/3, \ldots$ which converges to
zero. However $f(x_n) = 1, 1, 1, 1, \ldots$ converges to $1 \neq f(0) = 0$.

Another, perhaps more popular, way to define continuity does not
rely on sequences but uses the same idea that a function associates points
that are close in the range to points that are close in the domain. It uses
an $\epsilon > 0$ and absolute value to define closeness. It turns out that these
two definitions are equivalent.

$\epsilon - \delta$ **Continuity at a Point**   Let $f : D \rightarrow R$ be a real valued function of
real numbers. Let $x_0 \in D$. If for all $\epsilon > 0$ there exists a $\delta > 0$ such that
for all $x \in D$ we have that $|x - x_0| < \delta$ implies $|f(x) - f(x_0)| < \epsilon$ then f is
continuous at x_0.

So far our definition is just for one point which isn't very useful if we
want a function to be continuous on some interval or set.

Continuous on a Set   If $f : D \rightarrow R$ is continuous at every $c \in D$ then it is continuous on the set D. If D is the natural domain of the function, the domain where it is defined (say $x \geq 0$ for the function $\sqrt{x}$), then the function is said to be continuous.

Next is a series of examples that start out very short, and use the properties of sequences when doing so will make the proof short. Observe that by using the examples presented here one can build up an argument that polynomials are continuous. We also prove an interesting theorem about the continuity of the inverse of a monotonic function on an interval. This seemingly odd fact turns out to be very useful to establish the continuity of the nth root function. We use it in a later example for the cube root.

Constant Function is Continuous   Let $f(x) = k$, some real number. Then f is continuous.

Proof   Let $\epsilon > 0$, and $\delta = \epsilon$. Then if $|x - x_0| < \delta$ we have that $|k - k| < \epsilon$. Therefore f is continuous $\square$

Identity function is Continuous   Let $f(x) = x$. Then f is continuous.

Proof   Let $\epsilon > 0$, and $\delta = \epsilon$. Then $|x - x_0| < \delta$ implies $|x - x_0| < \epsilon$ and f is continuous $\square$

Sums are Continuous   Let f and g be real continuous functions from D to R. Then $f + g$ is continuous.

Proof   Let $x_n \rightarrow x$ in D. Then $f(x_n) \rightarrow f(x)$ and $g(x_n) \rightarrow g(x)$. By the sum of convergent sequences we have

$$(f(x_n) + g(x_n)) \rightarrow (f(x) + g(x)),$$

and $f + g$ is continuous $\square$

Products are Continuous   Let f and g be real continuous functions from D to R. Then fg is continuous.

Proof   Let $x_n \to x$ in D. Then $f(x_n) \to f(x)$ and $g(x_n) \to g(x)$. By the product of convergent sequences we have

$$(f(x_n)g(x_n)) \to (f(x)g(x)),$$

and fg is continuous $\square$

Composition is Continuous   Let $f : D \to R$ and $g : R \to Q$ be real continuous functions. Then the composition $g(f(x_0))$ is continuous at x_0.

Proof   Let $x_n \to x_0$ in D. Then $f(x_n) \to f(x_0)$. Since g is continuous at $f(x_0)$ we have that

$$g(f(x_n)) \to g(f(x_0)),$$

and the composition $g(f(x))$ is continuous $\square$

Monotonic Function   A function $f : D \to R$ is strictly monotonic increasing if whenever $x_0 < x_1$ we have that $f(x_0) < f(x_1)$. It is strictly monotonic decreasing if whenever $x_0 > x_1$ we have that $f(x_0) > f(x_1)$.

Inverse of Monotonic Function is Continuous   Let I be an open interval. If $f : D \to I$ is strictly monotonic onto I, then f^{-1} is continuous.

Discussion   Since we're going to prove continuity of the inverse, we want to conclude with

$$|f^{-1}(y) - f^{-1}(y_0)| < \epsilon.$$

Now lets unwrap this a little bit and see how we might use monotonicity:

$$-\epsilon < f^{-1}(y) - f^{-1}(y_0) < \epsilon,$$

$$f^{-1}(y_0) - \epsilon < f^{-1}(y) < f^{-1}(y_0) + \epsilon.$$

Since f is monotonic, we can apply it to each quantity in the inequality as follows

$$f(f^{-1}(y_0) - \epsilon) < f(f^{-1}(y)) < f(f^{-1}(y_0) + \epsilon),$$

and then rename the extreme quantities for convenience

$$y_1 < y < y_2.$$

Then since we're looking for a δ, split the distance between y_2 and y_1 at some point, say y_0, and see if we can get the inequalities to work out in the other direction from

$$|y - y_0| < \delta = \min\{y_2 - y_0, y_0 - y_1\}$$

Proof Let $x_1, x_0, x_2 \in D$ with $x_1 < x_0 < x_2$. Then by strict monotonicity we have

$$f(x_1) < f(x_0) < f(x_2)$$

and using the usual $y = f(x)$ notation we have

$$y_1 < y_0 < y_2.$$

Since f is strictly monotonic, the inverse f^{-1} is strictly monotonic as well and

$$f^{-1}(y_1) < f^{-1}(y_0) < f^{-1}(y_2).$$

Then for a sufficiently small $\epsilon > 0$, and since f^{-1} is defined on the interval I, There is a $y \in I$ such that

$$f^{-1}(y_0) - \epsilon < f^{-1}(y) < f^{-1}(y_0) + \epsilon.$$

Now let $\delta = \min\{y_2 - y_0, y_0 - y_1\}$. Then $|y - y_0| < \delta$ implies

$$|y - y_0| < y_2 - y_0 \text{ and } |y - y_0| < y_0 - y_1,$$

$$-y_2 + y_0 < y - y_0 < y_2 - y_0 \text{ and } -y_0 + y_1 < y - y_0 < y_0 - y_1,$$

$$y < y_2 \text{ and } y_1 < y.$$

Then we have

$$y_1 < y < y_2,$$

and by monotonicity

$$f^{-1}(y_1) < f^{-1}(y) < f^{-1}(y_2).$$

But then from above we know that

$$f^{-1}(y_0) - \epsilon < f^{-1}(y) < f^{-1}(y_0) + \epsilon,$$

$$-\epsilon < f^{-1}(y) - f^{-1}(y_0) < \epsilon,$$

$$|f^{-1}(y) - f^{-1}(y_0)| < \epsilon,$$

and f^{-1} is continuous $\square$

If our continuous function is going to be smooth, then we would expect that when the function acts on a closed and bounded set that the result is a closed and bounded set. If this didn't happen, then in some sense the function wouldn't be smooth enough to be called continuous. Next is an attempt to do such a proof using the sequential definition.

Boundedness of Continuous Functions   Let $f : [a, b] \to R$ be a real valued continuous function. Then there exists $M \in \mathbb{R}$ such that for all $x \in [a, b]$ we have $|f(x)| \le M$.

Discussion   Most authors give a proof by contradiction, but we'll try to avoid that technique by allowing the choice of an arbitrary, convergent, sequence from the closed interval. Indeed, for any $x_n \to x_0$ continuity says that $f(x_n) \to f(x_0)$ and by the definition of convergence $f(x_0)$ is a real number (not ∞ or $-\infty$). Recall as well that convergent sequences are bounded, and so there exists $M \in \mathbb{R}$ such that $f(x_n) \le M$.

Note that the closed interval is crucial because it contains the limit of any sequence from the interval. The open interval containing a convergent sequence might not contain the limit point. Since x_n is any sequence converging to x_0, the symbol x_n could be any value from the interval. Hence, we can drop the subscript at the end.

'Proof'   Let x_n be any sequence such that $x_n, x_0 \in [a, b]$ and $x_n \to x_0$. Then since f is continuous, $f(x_0)$ is a real number. Since $f(x_n)$ is convergent it is bounded, say by m. Let M be the larger of m or $f(x_0)$ so that $|f(x_n)| \le M$ for all $n \ge 0$. Since x_n is a convergent but otherwise arbitrary sequence from $[a, b]$, the bound holds for any $x \in [a, b]$, that is $|f(x)| \le M$.

The previous 'proof' has a flaw. It is the generalization from an arbitrary sequence in a closed interval to any element from the interval. It turns out that this is a valid generalization (in the space of real numbers with the usual topology) but it needs rigorous justification. The problem is that there are an uncountable number of real numbers and we need a mechanism to address all those cases.

To be more precise about what the problem is we need to consider how m was selected. It was based on a particular sequence. Another possible value in $[a, b]$ might have that $f(x_k) > m$ for some $x_k \in [a, b]$. In any interval there are a lot of real numbers to examine, and so we'll examine the formidable but necessary and rewarding machinery to cover those cases.

3.3   A Map of the Detour

Our usual tool of induction is not enough because there are more real numbers than there are natural numbers. Here is an outline of how we'll take care of this:

- First we'll make the case that almost each decimal number is a unique real number.

- Next we'll present the diagonal argument that there are more real numbers than natural numbers.

- Given that we need a small number of cases to cover all the real numbers in an interval, we'll use this idea as the definition of a new concept that is called compact.

- We'll review some of the properties of unions and intersections of sets when we apply a function to the sets, and then prove that a continuous function maps compact sets to compact sets.

- We'll prove that compact subsets of $\mathbb{R}$ are bounded and closed.

After this detour we'll pick back up with the properties of continuous functions on the real line.

3.4   Series and Decimals

A very common representation of a number is as a decimal, such as $3.1415926535...$ Lets take a closer look at what this notation means:

$$3+\frac{1}{10}+\frac{4}{100}+\frac{1}{1000}+\frac{5}{10000}+\frac{9}{100000}+\frac{2}{1000000}+\frac{6}{10000000}+\frac{5}{100000000}+\cdots$$

We can write this in a shorter form as

$$3 + \sum_{i=1}^{\infty} \frac{d_i}{(10)^i}$$

where $d_i \in \{0,1,2,3,4,5,6,7,8,9\}$. This is similar to a particular case of the geometric series with $r = 1/10$ but in our decimal the a is a different number on each term:

$$\sum_{i=0}^{\infty} a r^i.$$

Notice that if $r = 1$ we have

$$\sum_{i=0}^{\infty} a 1^i = a \sum_{i=0}^{\infty} 1 = a(1+1+1+1+\cdots)$$

which increases without bound, or diverges. To decide if a series converges we will look at the sequence of partial sums:

$$s_n = \sum_{i=0}^{n} a r^i.$$

For the geometric series with $r = 1$ the first few partial sums are

$$s_0 = a, \ s_1 = 2a, \ s_2 = 3a, \ s_3 = 4a, \text{ and so on: } s_n = (n+1)a.$$

For the case when $|r| < 1$ we'll write the terms and then multiply by r:

$$s_n = a + ar + ar^2 + \cdots + ar^{n-1}$$

$$r s_n = ar + ar^2 + ar^3 + \cdots + ar^n$$

Then form the difference $s_n - r s_n$,

$$s_n - r s_n = a - ar^n.$$

Now solve for s_n to get

$$s_n = \frac{a - ar^n}{1 - r}.$$

Then apply the limit laws for sequences to get

$$\lim_{n \to \infty} s_n = \lim_{n \to \infty} \frac{a - ar^n}{1 - r} = \frac{a}{1 - r} \lim_{n \to \infty} (1 - r^n) = \frac{a}{1 - r}.$$

Since the limit of the sequence of partial sums converges, the series converges when $|r| < 1$ and we have that

$$\sum_{i=0}^{\infty} ar^i = \frac{a}{1-r}.$$

Now lets consider the two numbers 1, and 0.9999999... Are they the same? Applying the geometric series says they are. Note that our series is missing the first term of the corresponding geometric series, and so it needs to be subtracted from the sum of the entire geometric series.

$$0.999999\cdots = 0 + \sum_{i=1}^{\infty} 9\left(\frac{1}{10}\right)^i = \frac{9}{1-(1/10)} - 9(1/10)^0 = 10 - 9 = 1.$$

On the previous page we used the fact that $\lim_{n\to\infty} r^n = 0$ for $|r| < 1$ without proof. We'll establish this fact using Bernoulli's Inequality.

Bernoulli's Inequality Let $d > -1$ and $n > 0$. Then

$$(1+d)^n \geq 1 + nd.$$

Proof Proceeding by induction, the base case of $n = 1$ yields

$$(1+d)^1 \geq 1 + 1d,$$

$$1 + d \geq 1 + d,$$

which is obviously true. Now for $n + 1$ we have

$$(1+d)^{n+1} = (1+d)^n(1+d).$$

Apply the induction assumption to get

$$(1+d)^{n+1} = (1+d)^n(1+d) \geq (1+nd)(1+d) = 1 + d + nd + nd^2 = 1 + (n+1)d + nd^2,$$

and since $nd^2 > 0$

$$(1+d)^{n+1} \geq 1 + (n+1)d.$$

Therefore, by the principle of mathematical induction the statement is true for all $n \geq 1$ $\square$.

Now we'll use this inequality in the proof that r^n converges when $|r| < 1$.

For $|r| < 1$, $r^n \to 0$: Discussion   We need to conclude that $|r|^n < \epsilon$, which is intuitively verifiable with a calculator, but not at all obvious how to prove. We have $(1+d)^n \geq 1+nd$ to use, but that doesn't appear to be much help at first. If we rewrite it as

$$\frac{1}{1+nd} \geq \frac{1}{(1+d)^n}$$

then we might try to make

$$\frac{1}{(1+d)^n} = |r|^n,$$

and we would be close. It turns out that the thing to do is to pick d so that

$$\frac{1}{(1+d)^n} = \frac{1}{|r|^n}.$$

Proof   Let $d = \frac{1}{|r|} - 1$. Then using Bernoulli's inequality we have

$$(1+d)^n \geq 1+nd,$$

$$\frac{1}{1+nd} \geq \frac{1}{(1+d)^n}$$

$$\frac{1}{1+nd} \geq \frac{1}{(1+\frac{1}{|r|}-1)^n}$$

$$\frac{1}{nd} \geq \frac{1}{(1+\frac{1}{|r|}-1)^n} = |r|^n \geq 0$$

Then take the limit as $n \to \infty$, and by the squeeze theorem we have

$$\lim_{n\to\infty} |r|^n = 0 \quad \square$$

Now that we established a useful tool, well return to the discussion on series.

Series A series is an infinite sum of the form

$$\sum_{k=m}^{\infty} a_k.$$

Convergent Series A series is said to converge when the limit of the sequence of partial sums converges as a limit to s, that is

$$\lim_{n\to\infty} \sum_{k=m}^{n} a_k = s.$$

Decimal Expansion Let $n, K \in \mathbb{N}$ and $d_n \in \{1,2,3,4,5,6,7,8,9,0\}$ then the list of digits (not a product)

$$K.d_1 d_2 d_3 d_4 \ldots$$

is a decimal expansion.

A Decimal Expansion is a Real Number Observe that

$$K.d_1 d_2 d_3 d_4 \cdots = K + \lim_{n\to\infty} \sum_{k=1}^{n} \frac{d_k}{10^k}.$$

Note that

$$\sum_{k=1}^{n} \frac{d_k}{10^k} \leq \sum_{k=1}^{n} \frac{9}{10^k},$$

and since the geometric series on the right converges we have

$$\lim_{n\to\infty} \sum_{k=1}^{n} \frac{d_k}{10^k} \leq \lim_{n\to\infty} \sum_{k=1}^{n} \frac{9}{10^k} = 1.$$

Since all the terms of the series are non-negative, the sequences of partial sums is monotonic. And since it is bounded (by one), we have a convergent sequence. Hence a decimal expansion is a real number □

We'll need an intermediate result to use to show that decimals are almost unique. This should be no surprise.

Series Inequality Lemma   Let a_k, b_k be nonnegative. If $a_k \leq b_k$ for all k
and if $a_k < b_k$ for one k, then if the series converge we have that

$$\sum_{k=m}^{\infty} a_k < \sum_{k=m}^{\infty} b_k.$$

Proof   Let $a = \sum a_k$ and $b = \sum b_k$. Then

$$b - a = \sum_{k=m}^{\infty} b_k - \sum_{k=m}^{\infty} a_k,$$

and replacing the summation notation with the sequence of partial sums
and applying limit laws yields

$$b - a = \lim_{n \to \infty} \sum_{k=m}^{n} b_k - \lim_{n \to \infty} \sum_{k=m}^{n} a_k$$

$$= \lim_{n \to \infty} \left(\sum_{k=m}^{n} b_k - \sum_{k=m}^{n} a_k \right) = \lim_{n \to \infty} \sum_{k=m}^{n} (b_k - a_k).$$

Now for each k we have $b_k \geq a_k$ and for at least one k we have $b_k > a_k$.
Therefore for at least one k we have $(b_k - a_k) \in \mathbb{R}_+$ and

$$b - a > 0, \quad \text{with} \quad \sum_{k=m}^{\infty} a_k < \sum_{k=m}^{\infty} b_k \; \square$$

Discussion   The next proof follows a structure similar to the one used
when we proved that the limit of a sequence is unique. We will suppose
that two distinct decimal expansions are the same real number and then
derive a contradiction. The algebra is a bit more intense, and we have
the special case that $1.0000\cdots = 0.9999\ldots$ to deal with as well. This makes
three categories that are addressed in two cases. This proof requires careful reading because all the details are not spelled out. But you've had
some practice so far, so give it a try. It might be useful to write the beginning of a decimal expansion and interpret the steps of the proof as they
would apply to your number.

Decimals Are Almost Unique Let $K.d_1 d_2 \ldots$ be a decimal expansion. Then exactly one of the following happens:

A: The decimal expansion ends in all nines, that is

$$K.d_1 d_2 \ldots d_r 9999999 \ldots$$

B: The decimal expansion ends in all zeroes, that is

$$K.d_1 d_2 \ldots (d_r + 1)0000000 \ldots$$

C: The decimal expansion is unique.

If A or B happen, then the decimal expansions are the same number.

Proof The case of A or B is left to the reader since it is a direct application of the geometric series that this section started with. For C, suppose that x can be written as

$$K.c_1 c_2 c_3 \ldots$$

or as

$$K.d_1 d_2 d_3 \ldots$$

Since the decimal expansions are different, let m be the first index with $c_j < d_j$ (just relabel if the digit of first difference has $d_j < c_j$). Now since we do not have repeating nines, by the series inequality lemma we have that

$$x = K.c_1 c_2 c_3 \cdots < K + \sum_{j=1}^{m} \frac{c_j}{10^j} + \sum_{j=m+1}^{\infty} \frac{9}{10^j}.$$

By applying the geometric series we have

$$x < K + \sum_{j=1}^{m} \frac{c_j}{10^j} + \sum_{j=m+1}^{\infty} \frac{9}{10^j} = K + \sum_{j=1}^{m} \frac{c_j}{10^j} + \frac{1}{10^m}.$$

Note that the digits are all the same up to and including the index $m-1$. In what we have written only $c_m < d_m$ so that

$$x < K + \sum_{j=1}^{m} \frac{c_j}{10^j} + \frac{1}{10^m} = K + \sum_{j=1}^{m-1} \frac{d_j}{10^j} + \frac{c_m}{10^m} + \frac{1}{10^m}.$$

Now observe that the right hand side is a truncation of the decimal expansion that uses the digits d_j. Also note that since $c_m < d_m$ we have that $c_m + 1 \leq d_m$, and so we have

$$x < K + \sum_{j=1}^{m-1} \frac{d_j}{10^j} + \frac{c_m + 1}{10^m} \leq K + \sum_{j=1}^{m} \frac{d_j}{10^j} \leq x.$$

The conclusion of the inequalities is that $x < x$, thus we have a contradiction. We must have been wrong when we started by supposing that x could be written as two distinct decimal expansions □

3.5 Countability

Now that we know a decimal expansion represents a real number, lets make a list of all the decimal expansions in $(0, 1)$ without repeating zeroes or nines.

$$0.654897351\ldots$$

$$0.329564815\ldots$$

$$0.149567423\ldots$$

$$0.874695328\ldots$$

$$0.129853478\ldots$$

$$\vdots$$

So we have an enumeration of the decimal numbers in a list. We can map each decimal to natural number since they are listed here in some ordering. If this holds then we say that the list is *countable*.

But if we go down the diagonal starting at the 6 in the top left and choose a digit from $\{1, 2, 3, 4, 5, 6, 7, 8\}$ that does not equal the observed digit on the diagonal, we can make a new decimal. It won't be on the list because the first digit differs from the first decimal, the second digit differs from the second decimal, and so on. Indeed $0.73812\ldots$ is not on the list. Therefore we must have an uncountable number of decimals. Since each decimal on the list is a real number, we have an uncountable number of real numbers. By *uncountable* we mean a collection of items than cannot be put in one-to-one correspondence with the natural numbers.

Recall that we are establishing the fact that induction is not strong enough to cover all the real numbers in a closed interval. So far we have seen that there are more real numbers than natural numbers. Since we can't list all the real numbers to align with the natural numbers, our usual induction doesn't apply.

For a contrast we'll demonstrate that the rational numbers are countable. We can make a table with all the integers on each axis. Then form the entries of the table like a multiplication table, but write the entries as rational numbers:

0	1	-1	2	-2	3 ...
1	1/1	1/-1	1/2	1/-2	1/3...
-1	-1/1	-1/-1	-1/2	-1/-2	-1/3...
2	2/1	2/-1	2/2	2/-2	2/3...
-2	-2/1	-2/-1	-2/2	-2/-2	-2/3...
3	3/1	3/-1	3/2	3/-2	3/3...
⋮	⋮	⋮	⋮	⋮	⋮

Then if we start in the top left corner and list the numbers as they appear along the diagonal from left column going up and to the right, then when we get to the top row we return on a parallel path just below the previous one we get a sequence like this:

$$0, 1, 1-1, 1, -1, 2, -1, -1, 2, -2, 2, 1, \frac{1}{2}, -2, 3, \frac{-1}{2}, \frac{-1}{2}, -2, -2, 3, \ldots$$

which is a sequence which includes every rational number. We know it does because the numerators come from the list of integers on the top row, and the denominators come from the list of integers on the left column, and every combination occurs. This list is not very efficient because it includes many repetitions. A formula with no repetitions is the Calkin- Wilf Sequence given by:

$$q_{n+1} = \frac{1}{2\lfloor q_n \rfloor - q_n + 1}$$

Where $\lfloor q \rfloor$ is the largest integer less than or equal to q.

Every open set on the real line, $U \subset \mathbb{R}$, is a finite or countable union of open intervals, (a_n, b_n)

Discussion   The initial insight to get started is that every open interval contains a rational number, and that the rational numbers are countable. Then the procedure is to build open sets around the rational numbers until they coincide with U. If q is a rational number in U, and U is open, then there is some open interval around q but in U. To account for the fact that q might not be in the center we have $(a, q]$ and $[q, b)$ whose union is an open interval. Then we can choose a and b by using the infimum and supremum. Finally we verify that the union of all the open intervals is indeed U. If in the process of reading this proof it appears that we are leaving steps out, the reader should go back to the definitions of a_n and b_n.

Proof Let U be an open subset of $\mathbb{R}$. Since open sets contain rationals, let q_n be an enumeration of the rationals in U. Define two sequences as follows:

$$a_n = \inf\{a \mid (a, q_n] \subset U\},$$
$$b_n = \sup\{b \mid [q_n, b) \subset U\}.$$

Then we want to show that

$$U = \bigcup_{n=1}^{\infty}(a_n, b_n).$$

Let $x \in U$. Then for each n we have three cases:

C1 If $x = q_n$ then $x \in (a_n, b_n)$

C2 If $x < q_n$ then $x \in \{a \mid (a, q_n] \subset U\}$, and $x \geq a_n$ so that $x \in (a_n, b_n)$.

C3 If $x > q_n$ then $x \in \{b \mid [q_n, b) \subset U\}$, and $x \leq b_n$ so that $x \in (a_n, b_n)$.

Therefore

$$U \subset \bigcup_{n=1}^{\infty}(a_n, b_n).$$

Now let

$$x \in \bigcup_{n=1}^{\infty}(a_n, b_n).$$

Then there is at least one interval, i such that $x \in (a_i, b_i)$. So $x \in (a, q_i] \cup [q_i, b)$ and $x \in U$. So that

$$\bigcup_{n=1}^{\infty}(a_n, b_n) \subset U$$

and then

$$\bigcup_{n=1}^{\infty}(a_n, b_n) = U \ \square$$

3.6   Compact sets

To get ready for working with functions of sets, lets recall a few facts about how functions and sets relate to each other. In what follows we take for granted the existence of the inverse function. Note that the inverse function doesn't always exist, in which case these results don't hold.

Image   Given a function $f : A \rightarrow B$. Let $X \subset A$. Then $f(X)$ is the image of f acting on X. It is the set

$$\{f(x)|x \in X\}$$

Pre-image   Given a function $f : A \rightarrow B$. Let $Y \subset B$. Then when the inverse exists, $f^{-1}(Y)$ is the pre-image of Y for the function f. It is the set

$$\{f^{-1}(y)|y \in Y\}$$

Pre-image union   The pre-image of the union of two sets is the union of the pre-images, that is

$$f^{-1}(A \cup B) = f^{-1}(A) \cup f^{-1}(B).$$

Proof   Let $x \in f^{-1}(A \cup B)$. Then $f(x) \in A \cup B$. So $f(x) \in A$ or $f(x) \in B$ and we have $x \in f^{-1}(A)$ or $x \in f^{-1}(B)$. Hence $x \in f^{-1}(A) \cup f^{-1}(B)$.
For the other direction suppose $x \in f^{-1}(A) \cup f^{-1}(B)$. Then $x \in f^{-1}(A)$ or $x \in f^{-1}(B)$. Hence $f(x) \in A$ or $f(x) \in B$, and $f(x) \in A \cup B$ which means that $x \in f^{-1}(A \cup B)$ $_\square$

Pre-image intersection   The pre-image of the intersection of two sets is the intersection of the pre-images, that is

$$f^{-1}(A \cap B) = f^{-1}(A) \cap f^{-1}(B).$$

Proof   Let $x \in f^{-1}(A \cap B)$. Then $f(x) \in A \cap B$ so that $f(x) \in A$ and $f(x) \in B$. Therefore $x \in f^{-1}(A)$ and $x \in f^{-1}(B)$. Hence $x \in f^{-1}(A) \cap f^{-1}(B)$.
For the other direction suppose $x \in f^{-1}(A) \cap f^{-1}(B)$. Then $x \in f^{-1}(A)$ and $x \in f^{-1}(B)$. And therefore $f(x) \in A$ and $f(x) \in B$ so that $f(x) \in A \cap B$. Hence $x \in f^{-1}(A \cap B)_\square$

Image union   The image of the union of two sets is the union of the images, that is

$$f(A \cup B) = f(A) \cup f(B).$$

Proof   Let $y \in f(A \cup B)$. Then there is an $x \in A \cup B$ with $y = f(x)$. So $x \in A$ or $x \in B$ and therefore $f(x) \in f(A)$ or $f(x) \in f(B)$. Hence $y = f(x) \in f(A) \cup f(B)$.
For the other direction let $y = f(x) \in f(A) \cup f(B)$. Then $f(x) \in f(A)$ or $f(x) \in f(B)$. So $x \in A$ or $x \in B$ and $x \in A \cup B$. Therefore $y = f(x) \in f(A \cup B)_\square$

Image intersection   The image of the intersection of two sets is a subset of the intersection of their images, that is

$$f(A \cap B) \subset f(A) \cap f(B)$$

Proof   Let $y \in f(A \cap B)$. Then there is $x \in A \cap B$ such that $f(x) = y$. So $x \in A$ and $x \in B$ and therefore $f(x) \in f(A)$ and $f(x) \in f(B)$. Hence $f(x) = y \in f(A) \cap f(B)_\square$

Note that the above results all hold for arbitrary intersections and unions. It would be a good exercise to come up with a counter example that shows why we do not have equality for the image intersection property. Hint: think in at least two dimensions.

Now there are only two more definitions before compactness.

Open cover   An open cover of a set X is a collection of open sets, $\{G_\alpha\}$ such that

$$X \subset \bigcup_\alpha G_\alpha.$$

The greek subscript indicates that the union may be of an uncountable number of sets.

Finite sub cover   A finite sub cover of an open cover $\{G_\alpha\}$ is a finite collection, $\{G_1, G_2, \dots G_n\}$ that covers X.

Compact set   A set is compact is every open cover has a finite sub cover.

We'll make the definition of continuity that we started the chapter with formal. Recall that it was in the body of the third paragraph of the chapter.

Continuous function   Let $f : X \to Y$ be a function. If for every open set $U \subset Y$ we have that $f^{-1}(U)$ is open in X then f is continuous. In other words, the pre-image of any open set is open.

Observe how the idea of a set being compact makes the proof of the next theorem very elegant:

Continuous functions preserve compactness   Let $f : X \to Y$ be continuous and $K \subset X$ be compact. Then $f(K)$ is compact.

Proof   Let $f : X \to Y$ be continuous and $K \subset X$ be compact. Then suppose $\{G_\alpha\}$ is an open cover of $f(K)$. Now

$$\{f^{-1}(G) | G \in \{G_\alpha\}\}$$

is a collection of sets covering K (by the pre-image union property). Since f is continuous they are open sets. Since K is compact, a finite number of them cover X, say

$$\{f^{-1}(G_1), f^{-1}(G_2), f^{-1}(G_3) \dots f^{-1}(G_n)\},$$

and then by pre-image union $\{G_1, G_2, G_3 \dots G_n\}$ cover $f(K)$ $\square$

Ok, thats great about these compact sets. It is nice that continuous functions preserve that property. But what if there aren't any compact sets, or they're all empty. That doesn't do us any good. Well, it turns out that there are lots of compact sets. We'll start in the easier direction and show that compact sets are bounded, indeed this is where we left off with sequences and continuity. Then we'll do the rest of the work for what is commonly known as the Heine-Borel Theorem. It applies to $\mathbb{R}^n$ but we will stay with just $\mathbb{R}$.

Heine-Borel theorem A set of real numbers is compact if and only if it is closed and bounded.

We'll prove all the parts of this in smaller chunks. So first is

Compact $\Longrightarrow$ bounded Let $X \subset \mathbb{R}$ be compact. Then X is bounded.

Discussion The insight needed here is how to come up with the initial covering. Since we have a compact set it has to have an open covering. If it didn't we wouldn't have a compact set. Then we're not interested in the trivial empty set, so there is some element in the set. Since we're restricted to the real line, we can always pick an integer larger than some real number in the set. We use this idea to construct the covering.

Proof Let $x_0 \in X$. Then $\{(x_0 - n, x_0 + n)|n \in \mathbb{N}\}$ is an open cover of X. Since X is compact there is a finite sub cover

$$\{(x_0 - k_1, x_0 + k_1), (x_0 - k_2, x_0 + k_2), \ldots (x_0 - k_p, x_0 + k_p)\}.$$

Really only one of these intervals is necessary since when the k_i are increasing they are nested. So let k_p be the largest and by the definition of cover

$$X \subset (x_0 - k_p, x_0 + k_p).$$

Let $x \in X$. Then $x \in (x_0 - k_p, x_0 + k_p)$ and $|x| \leq \max\{x_0 - k_p, x_0 + k_p\}$. Therefore X is bounded $\square$

Compact $\Longrightarrow$ closed   Let $X \subset \mathbb{R}$ be compact. Then X is closed.

Discussion   Since there is no ordering of the points in the plane, this proof is usually done by contradiction because the higher dimension case is considerably different. We will exploit the ordering of the real numbers. Since compactness is defined as a property for *every* open cover, it is rare to see a direct proof that a set is compact. On the other hand if we can construct an open cover of a set that has no finite sub cover, then we know the set is not compact. If we prove the contrapositive, that will be the conclusion we need, and so that is the approach we use here.

Proof   We will prove the contrapositive, if X is not closed then X is not compact. Since X is not closed, there is a convergent sequence x_n from X converging to a point x with $x \notin X$. By the order properties there are three cases:

- Case 1: For all $a \in X$, $a > x$.

- Case 2: For all $b \in X$, $b < x$.

- Case 3: X is the union of two sets separated by x:

$$A = \{a | a > x \text{ and } a \in X\}$$

and

$$B = \{b | b < x \text{ and } b \in X\}.$$

Case 3 is transformed into Case 1 or 2 in the following way. Define a subsequence of x_n by selecting the terms of x_n in the following way. Here we interpret x_i as each number in the sequence x_n iteratively:

$$\text{If } x_i \in A \text{ Then } x_{n_k} = x_i$$

or

$$\text{If } x_i \in B \text{ Then } x_{n_l} = x_i.$$

Then x_{n_k} or x_{n_l} converges to x (one may be a finite list that is irrelevant). As a result A falls in case 1 and B falls in case 2. Since A and B are disjoint these two cases are all we need. Now we will do case 1. The only change for case two is some inequality signs and subtraction operations so it is omitted.

- Case1: For all $a \in X$, $a > x$.

We construct a cover for X and use the property that X is not closed to show there is no finite sub cover of our cover. Then conclude that X is not compact. We'll first prove that

$$\bigcup_{n \in \mathbb{N}} (x_n, n).$$

is a cover for X. That is we show that

$$X \subset \bigcup_{n \in \mathbb{N}} (x_n, n).$$

Let $y \in X$. Then $y > x$ and $y - x = \epsilon > 0$. By convergence there exists $N \in \mathbb{N}$ such that $n > N$ implies

$$|x_n - x| < y - x = \epsilon,$$

$$x_n - x < y - x,$$

$$x_n < y,$$

and so $y \in (x_n, n)$. Therefore

$$y \in \bigcup_{n \in \mathbb{N}} (x_n, n),$$

and our cover actually covers X. Next we demonstrate how to find $x_i \in X$ that is not in any finite sub collection of $\{(x_n, n)\}$ Let n_k be the selection function, that is

$$\{(x_{n_1}, n_1), (x_{n_2}, n_2), (x_{n_3}, n_3), \ldots (x_{n_M}, n_M)\}$$

is some finite sub cover. Then define ϵ as follows:

$$\epsilon = \min\{(x_{n_1} - x), (x_{n_2} - x), (x_{n_3} - x), \ldots (x_{n_M} - x)\}.$$

By convergence of x_n there exists $N \in \mathbb{N}$ such that $k > N$ implies

$$|x_k - x| < \epsilon,$$

Which means that x_k is not covered. So we have exhibited an open cover of X that has no finite sub cover. Therefore X is not compact.

Recall that we are proving the contrapositive. In order to show that compact sets are closed, we supposed that X was not closed and concluded that X is not compact. The equivalent statement is that compact sets are closed $\square$

Closed and bounded $\implies$ compact   Let $[a, b]$ be a closed and bounded subset of $\mathbb{R}$. Then $[a, b]$ is compact.

Proof   We will assume that $[a, b]$ is not compact and derive a contradiction. Let O be an open cover of $[a, b]$. Since $[a, b]$ is not compact, either

$$\left[a, \frac{a+b}{2}\right] \text{ or } \left[\frac{a+b}{2}, b\right]$$

does not have a finite sub cover. Choose one without a finite sub cover (say the left side) and divide it again to get

$$\left[a, \frac{a+b}{4},\right] \text{ and } \left[\frac{a+b}{4}, \frac{a+b}{2}\right].$$

At least one of which does not have a finite sub cover. Repeating this process yields a sequence of nested intervals with the following three properties

1. $[a_{n+1}, b_{n+1}] \subset [a_n, b_n]$.

2. $b_n - a_n = \frac{b-a}{2^n}$.

3. $[a_n, b_n]$ does not have a finite sub cover.

Since the closed intervals are nested (property 1) the intersection of the intervals is not empty and there exists x such that

$$x \in \bigcap_{n \in \mathbb{N}} [a_n, b_n].$$

By containment (from the section on open and closed sets) there are intervals so that

$$x \in (x - \epsilon, x + \epsilon) \subset (c, d) \subset O.$$

Now choose N so that $\frac{b-a}{2^N} < \epsilon$ and we have that $x \in [a_N, b_N]$. But then

$$[a_N, b_N] \subset (c, d) \subset O$$

which contradicts our initial assumption that a finite sub cover does not exist $\square$

Continuous functions do not preserve openness   Consider the function $f(x) = x^2$ acting on $(-1, 1)$. We have that $f((-1, 1)) = [0, 1)$ which is not open.

Chapter 4

Continuation of Continuity

In this chapter we will show the equivalence of the several definitions of continuity that we have presented so far. After that we will present some of the common properties of continuous functions, and then at the end of the chapter we'll point out the limitation of continuity as a link to the stronger property of uniform continuity. We will continue to use the fact that open sets consist of unions of basis elements (a, b), and that the containment principle says any particular non-empty basis element contains many more.

Continuous function   Let $f : X \to Y$ be a function. If for every open set $U \subset Y$ we have that $f^{-1}(U)$ is open in X then f is continuous. In other words, the pre-image of any open set is open.

$\epsilon - \delta$ Continuity at a point   Let $f : D \to R$ be a real valued function of real numbers. Let $x_0 \in D$. If for all $\epsilon > 0$ there exists a $\delta > 0$ such that for all $x \in D$ we have that $|x - x_0| < \delta$ implies $|f(x) - f(x_0)| < \epsilon$ then f is continuous at x_0.

Sequential continuity at a point   Let $f : D \to R$ be a real valued function of real numbers and x_n any sequence converging to $x_0 \in D$. Then if $f(x_n) \to f(x_0)$ for *every* sequence converging to x_0, f is continuous at x_0.

4.1   Equivalent Definitions of Continuity

Discussion   We use the fact that

$$x_0 \in (a, b) \text{ implies } |x - x_0| < \epsilon.$$

That is

$$x_0 \in (a, b) \Longrightarrow \exists \, \epsilon \le \min\{x_0 - a, b - x_0\} \text{ so that } |x - x_0| < \epsilon.$$

Continuity $\Longrightarrow$ $\epsilon - \delta$ continuity   Let $f : D \to R$ be continuous by open sets. Then f is continuous in $\epsilon - \delta$ definition.

Proof   Let $f(x_0)$ be a point in an open set, U in the range of f, that is

$$f(x_0) \in U \subset R.$$

Then there is a basis element so that $f(x_0) \in (a, b) \subset R$. Note that $f(x_0)$ represents any particular point in the range of f and (a, b) is any basis element containing $f(x_0)$. Then let $\epsilon \le \min\{f(x_0) - a, b - f(x_0)\}$, and ϵ is any positive value (because it is associated with the basis element which could be any basis element in the range). Since

$$f(x_0) \in (a, b),$$

we have that

$$f^{-1}(f(x_0)) \in f^{-1}((a, b)),$$

and so

$$x_0 \in f^{-1}((a, b)).$$

By the open set definition of continuity, $f^{-1}((a, b))$ is an open set, and therefore is a union of basis elements. Choose one containing x_0 and label it (c, d) so that

$$x_0 \in (c, d) \subset D.$$

Define $\delta = \min\{x_0 - c, d - x_0\}$ and then $\delta > 0$ and depends upon the value of ϵ. Now we write what we have done in terms of absolute value, consistent with the open intervals given by the continuous function and basis elements.

So for any ϵ we have found a δ such that whenever $|x - x_0| < \delta$ we have $|f(x) - f(x_0)| < \epsilon$ $\square$

$\epsilon - \delta$ **continuity** $\implies$ **continuity** Let $f : D \to R$ be continuous by the $\epsilon - \delta$ definition. Then f is continuous by the open set definition.

Discussion Note that we can translate from absolute value and inequality to intervals like this:

$$|x - x_0| < \delta \implies x \in (x_0 - \delta, x_0 + \delta), \text{ a basis element.}$$

Also, as a reminder, x_0 is an arbitrary, but fixed point in the domain while x is not fixed. Note that the $\epsilon - \delta$ definition says that there exists a δ for each ϵ, and this δ determines a basis element.

Proof Since f is $\epsilon - \delta$ continuous, for any $\epsilon > 0$ we have

$$|f(x) - f(x_0)| < \epsilon,$$

which means that $f(x) \in (a, b)$. Where (a, b) is a basis element dependent on ϵ, containing $f(x_0)$. Note that one may add any other basis element to form an open set U containing $f(x_0)$ and not affect the rest of the proof. Then our definition says there exists a δ such that

$$|x - x_0| < \delta$$

which means that $x \in (c, d)$. Where (c,d) depends on δ. Next note that for any x such that

$$f(x) \in (a, b)$$

we have that

$$f^{-1}(f(x)) \in f^{-1}((a, b))$$

and

$$x \in f^{-1}((a, b)).$$

But also, the same x is such that $x \in (c, d)$, an open set in the domain. Therefore the pre-image of an open set $f^{-1}((a, b))$, is an open set (c, d) $\square$

$\epsilon - \delta$ **continuity** $\implies$ **sequential continuity**   Let $f : D \to R$ be continuous by the $\epsilon - \delta$ definition. Then f is continuous by the sequence definition.

Proof   Let x_n be a sequence converging to x_0. Then by convergence, for all $\delta > 0$ there exists N so that $n > N$ implies

$$|x_n - x_0| < \delta$$

But by $\epsilon - \delta$ continuity, for all $\epsilon > 0$ we have that $|x_n - x_0| < \delta$ implies

$$|f(x_n) - f(x_0)| < \epsilon.$$

So $f(x_n)$ converges to $f(x_0)_\square$

Sequential continuity $\implies$ $\epsilon - \delta$ **continuity**   Let $f : D \to R$ be continuous by the sequence definition. Then f is continuous by the $\epsilon - \delta$ definition.

Proof   We will prove the equivalent, contrapositive statement. That is, If a function is not $\epsilon - \delta$ continuous then it is not sequentially continuous. The next two lines are the $\epsilon - \delta$ definition and then its negation (our starting point).

$$\forall \, \epsilon > 0 \, \exists \, \delta > 0 \text{ such that } |x - x_0| < \delta \implies |f(x) - f(x_0)| < \epsilon$$

$$\exists \, \epsilon > 0 \, \forall \, \delta > 0 \text{ such that } |x - x_0| < \delta \implies |f(x) - f(x_0)| \geq \epsilon$$

Let x_n be a sequence converging to x_0 Then for all $\delta > 0$ there exists $N \in \mathbb{N}$ such that $n > N$ implies $|x_n - x_0| < \delta$. And by the negation of continuity there is an $\epsilon > 0$ such that

$$|f(x_n) - f(x_0)| \geq \epsilon$$

which means that $f(x_n)$ does not converge to $f(x_0)$. Now the contrapositive must also be true; if f is continuous by the sequential definition it is continuous by the $\epsilon - \delta$ definition $_\square$

4.2 Intermediate and Extreme Value Theorems

If you had a not very good calculus teacher he might have told you that a continuous function is one that you can draw without picking up your pencil. The problem with such a notion is that $\sin\left(\frac{1}{x}\right)$ on $(0,1)$ is continuous but rather hard to draw with any precision, even for your graphing calculator or computer algebra system. The property that inspires such a description is the intermediate value property. It says that if a continuous function takes on two values, it takes on every value in between. The proof is a combination of continuity with the supremum of a set defined by some point in the middle of the two values.

Here we state the theorem with strict inequalities, that is $f(a) < y < f(b)$, but non strict inequalities would work as well. We would then have two additional cases to consider, $y = f(a)$ and $y = f(b)$.

Also note that the extreme value theorem is often presented before the intermediate value theorem. I find that most proofs of the extreme value theorem implicitly use the intermediate value theorem and so I present the intermediate value theorem first.

Recall the problem of finding the root for $\sqrt{x+1} - \sqrt[3]{x} - .5 = 0$ in the section on the secant method. With the intermediate value theorem we can prove that the solution exists and give bounds on its value. We'll do this in an example after we prove the theorems.

Intermediate value theorem   Let $f : D \rightarrow R$ be a continuous function. If there exists y such that $f(a) < y < f(b)$ or $f(b) < y < f(a)$, then there exists $c \in (a, b)$ such that $f(c) = y$.

Discussion   We return to using sequences in most of our proofs. The insight for this proof is to construct a set where c is the supremum. Then by completeness it exists, and then we use inequalities to to conclude $f(c) = y$ as is required. In order to do that we'll make a sequence on either side of c. By the squeeze lemma and continuity we'll get what we want.

Proof   Consider the points from the domain where $f(x)$ is less than y:

$$S = \{x \in (a, b) \mid f(x) < y\}.$$

It is not empty and is bounded above by b so it has a supremum. Let

$$c = \sup\{x \in (a, b) \mid f(x) < y\}.$$

Note that $c - \frac{1}{n}$ is not an upper bound, and so

$$c - \frac{1}{n} \le c,$$

and there are s_n such that

$$c - \frac{1}{n} \le s_n \le c.$$

By the squeeze lemma $s_n \rightarrow c$. Since f is continuous $f(s_n) \rightarrow f(c) \le y$. Now for the other side let t_n be such that

$$c \le t_n \le c + \frac{1}{n},$$

and by the squeeze lemma $t_n \rightarrow c$. Since $t_n \notin S$ and $f(t_n) \ge y$, by continuity $f(t_n) \rightarrow f(c) \ge y$. Since $f(c) \ge y$ and $f(c) \le y$, we have that $f(c) = y$ □

Extreme value theorem Let $f : [a, b] \to R$ be a continuous function. Then there exists $c, d \in [a, b]$ such that for all $x \in [a, b]$ we have that $f(c) \le f(x) \le f(d)$.

Discussion In this proof we initially use the fact that continuity preserves compactness to conclude that $f([a, b])$ is bounded. Then we make a sequence $f(x_n)$ converging to the least upper bound. We are tempted to make the case that $x_n \to d$ (and that $d \in [a, b]$) directly by applying a definition, but this doesn't work out very well. Arguing indirectly through the use of subsequences, continuity, and the fact that a subsequence of a convergent sequence converges to the same point gets us where we need to go.

Proof Since f is continuous and $[a, b]$ is closed and bounded, we have that $f([a, b])$ is closed and bounded. Let

$$M = \sup\{f(x) \mid x \in [a, b]\}.$$

Then by the intermediate value theorem there exists $x_n \in [a, b]$ so that

$$M - \frac{1}{n} \le f(x_n) \le M.$$

By the squeeze lemma $f(x_n) \to M$. Since $x_n \in [a, b]$ and $[a, b]$ is bounded, there is a convergent subsequence x_{n_k} converging to some value, say d. Since $[a, b]$ is closed, $d \in [a, b]$. Then by continuity $f(x_{n_k}) \to f(d)$. Since $f(x_{n_k})$ is a convergent subsequence of the convergent sequence $f(x_n)$, they must converge to the same point. Hence $f(x_n) \to f(d)$. Finally, since $f([a, b])$ is closed, $f(d) \in f([a, b])$ and the supremum is attained by f at $d_\square$

Next we'll reach back to the theorem that a monotonic function onto an interval has a continuous inverse. We'll use that theorem to prove that the cube root is continuous. Most of the work will be to show that $f(x) = x^3$ is monotonic. It's very easy to visualize the function and conclude that it is monotonic, but not very rigorous. A similar technique works for any other root function.

The cube root is continuous   Let $f : \mathbb{R} \to \mathbb{R}$ be $f(x) = \sqrt[3]{x}$. Then f is continuous.

Proof   Let $g(x) = x^3$. We show that $g(x)$ is monotonic, then show that it is continuous so that by the intermediate value theorem its image is an interval. Finally we apply the fact that the inverse of a monotonic function onto an interval is continuous in order to get that $\sqrt[3]{x}$ is continuous. Let $x < y$, then by example 11 and some algebra

$$(y - x)^2 x < y(y - x)^2,$$

$$(y^2 - 2xy + x^2)x < (y^2 - 2xy + x^2)y,$$

$$3y^2 x - 3x^2 y + x^3 < y^3,$$

$$3xy(y - x) + x^3 < y^3.$$

Then if x, y are both positive or both negative we have

$$x^3 < 3xy(y - x) + x^3 < y^3,$$

and $x^3 < y^3$. On the other hand if $x < 0 < y$ then by example 11, $x^3 < 0$ and $0 < y^3$ so that $x^3 < y^3$ and in any case $g(x) = x^3$ is monotonic. To show that g is continuous, recall that the identity function, $h(x) = x$ is continuous, and that products of continuous functions are continuous. So $h(x)h(x)h(x) = x^3$ is continuous. Then by the intermediate value theorem the image of $g(x)$ is an interval. Finally, since $g(x) = x^3$ is monotonic on an interval, its inverse, $f(x) = \sqrt[3]{x}$ is continuous $\square$

At this point one can see how to apply this technique to any other root function. From now on we'll use the fact that sums, products, and compositions of polynomials and roots are continuous. However, it would be good practice to formulate a proof by induction that polynomials are continuous.

Next we'll apply the intermediate value theorem to prove that the root of

$$\sqrt{x + 1} - \sqrt[3]{x} - .5 = 0$$

actually exists. Recall the last example from the section on the secant method didn't appear to be converging to a solution. This technique will tell us an interval where we should expect the iterations to converge.

Example 32 The equation $\sqrt{x+1}-\sqrt[3]{x}-0.5=0$ has a real root.

Proof We need to establish the givens for the intermediate value theorem. Let
$$f(x)=\sqrt{x+1}-\sqrt[3]{x}-0.5.$$
Since $f(x)$ is the sum and composition of polynomials and roots on $(-1,\infty)$ it is continuous on $(-1,\infty)$. Then by guess and check we find two values of x, one that evaluates to a negative number and one that evaluates to a positive number. Hence, $f(0)=1/2$ and $f(1)=\sqrt{2}-3/2<0$. Then by the intermediate value theorem there exists $c\in(0,1)$ such that $f(c)=0$ $\square$

Example 33 Prove $x^7+x^5=1$ has a solution.

Discussion Note that for quadratic polynomials we have a formula to find the root. For higher order polynomials there is sometimes a formula, but not always. There are also situations involving trigonometric equations like $\cos x = x$ or the exponential function as in $xe^x = 5$ where the rules of algebra are not sufficient to find a root. To understand why, take a course in abstract algebra. Nevertheless our computers and calculators will tell us what the root is in most cases. A case that doesn't work is $x^2+1=0$. It has no real roots. It would be good to identify why the intermediate value theorem doesn't apply to this situation.

Proof Let $f(x)=x^7+x^5-1$ which is continuous because it is a polynomial. Note that $f(0)=-1$ and $f(1)=1$. Then by the intermediate value theorem there exists $c\in(0,1)$ such that $f(c)=0$ $\square$

Recall that before sequences and continuity we did a rather intricate proof to conclude that $\sqrt{x}$ was actually a real number. Now it just takes a few lines:

$\sqrt{b}$ is a real number Let $b>0$ and $x^2=b$. Let $f(x)=x^2-b$. Then $f(0)<0$ and $f(b+1)=b^2+2b+1-b>0$. By the intermediate value theorem there exists $c\in(0,b+1)$ such that $f(c)=0$, hence $c=\sqrt{b}$ is real $\square$

The next application of the intermediate value theorem illustrates its usefulness. We'll start with a function that is continuous but otherwise undefined, where we know that the domain and range are the same interval. From just this little bit of information we can conclude that the function has a fixed point, that there is an x_0 in the domain with $f(x_0) = x_0$. We will rely on the fact that the intermediate value theorem is true for closed intervals, even though we only proved it for open intervals.

Fixed Point Theorem   Let $f : [0,1] \to [0,1]$ be a continuous function. Then there exists $x_0 \in [0,1]$ such that $f(x_0) = x_0$.

Proof   Let $g(x) = f(x) - x$ and it is continuous because it is the difference of two continuous functions. Next observe that due to the restrictions on the range of f we have that

$$g(0) = f(0) - 0 = f(0) \geq 0,$$

and

$$g(1) = f(1) - 1 \leq 0.$$

Then we have that

$$g(1) \leq 0 \leq g(0)$$

and by the intermediate value theorem there exists $x_0 \in [0,1]$ such that $g(x_0) = 0$ and so

$$g(x_0) = f(x_0) - x_0 = 0,$$

$$f(x_0) = x_0 \; \square$$

At this point a natural direction to go would be to introduce the derivative and discuss how it is used to find maximum and minimum values. On the other hand, that topic is covered in most calculus books in some depth, although usually in such a manner that all the exercises are using continuous functions where the student just mechanically follows the process.

Rather than repeat that material (in greater depth) we will transition to the concept of uniform continuity. Our naive understanding of the Riemann integral implicitly uses the idea of uniform continuity, so we'll start to explore what that means.

This page intentionally left blank.

4.3   Uniform Continuity

The difference between a continuous function and a uniformly continuous function when you read the definitions for the first time is not usually clear. The definitions appear very similar at first glance. For this reason we'll explore some of the differences between the two ideas by highlighting them in some bullet statements.

- Uniform continuity is defined on a set.

- Continuity is defined at a point, even when a function is continuous at each point in an interval.

- With uniform continuity, there is a choice of δ that only depends on ϵ.

- With continuity, δ depends on ϵ and may also depend on the point x_0.

- If a function is uniformly continuous on a set, it is continuous at each point in the set.

- If a function is continuous at each point in a set, it is *not* necessarily uniformly continuous on the set.

- The statement, "uniformly continuous at a point" is nonsensical.

Uniformly continuous   Let $f : D \to R$ be a real valued function of real numbers. Then f is uniformly continuous on $X \subset D$ if for all $\epsilon > 0$ there exists $\delta > 0$ such that $x, y \in D$ and $|x - y| < \delta$ implies $|f(x) - f(y)| < \epsilon$.

$\epsilon - \delta$ Continuity at a point   Let $f : D \to R$ be a real valued function of real numbers. Let $x_0 \in D$. If for all $\epsilon > 0$ there exists a $\delta > 0$ such that for all $x \in D$ we have that $|x - x_0| < \delta$ implies $|f(x) - f(x_0)| < \epsilon$ then f is continuous at x_0.

A classic description to illustrate how δ can depend on the point of continuity is the function $1/x$ plotted on the next page. We'll think about what happens as the two bold dots follow the function to the left, but keeping ϵ fixed. Keep in mind the order in the definition of continuity.

First we fix ϵ, then we find a δ that implies $|f(x)-f(x_0)| < \epsilon$. The δ segment shown in the picture is the largest possible for the shown ϵ. Observe that as the point of continuity, x_0, moves toward zero, that δ must decrease. This illustrates how δ depends on the point x_0.

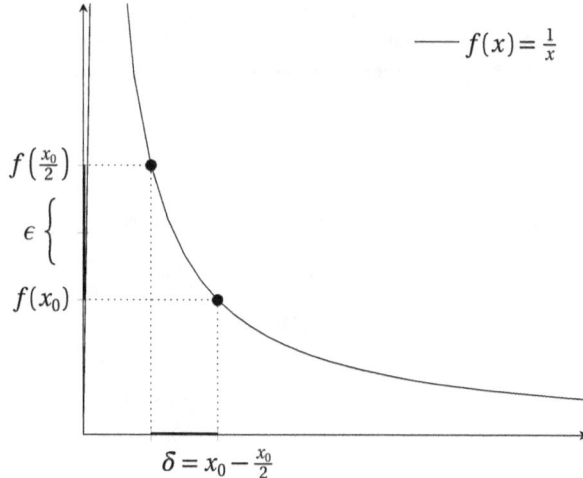

Now with x_0 moved to the left we see how δ must be smaller.

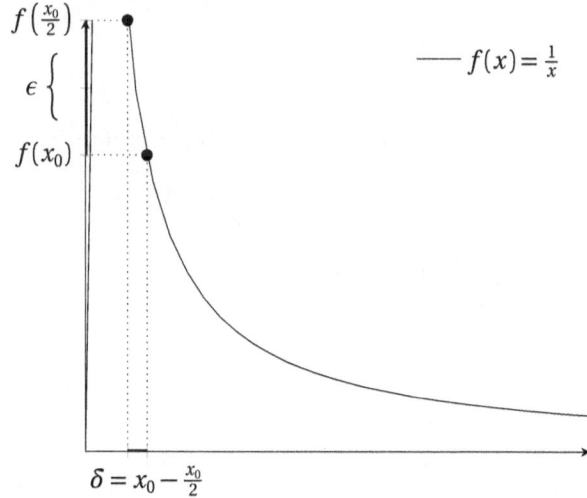

Next we'll examine some continuous functions on some given sets. Pay attention to the boundedness of the domain, and think about how steep the function is at different points in relation to whether the function is uniformly continuous or not. Note that in the definition for uniformly continuous we use y instead of x_0 to emphasize that it is a property of a function on a set, not at a point.

- $f(x) = x$ is uniformly continuous on $(-\infty, \infty)$.

- $f(x) = x^2$ is not uniformly continuous on $(-\infty, \infty)$.

- Let $b \in \mathbb{R}$, then $f(x) = x^2$ is uniformly continuous on $(-b, b)$.

- $f(x) = 1/x$ is not uniformly continuous on $(0, 1]$.

- $f(x) = \sqrt{x}$ is uniformly continuous on $(0, 1)$.

We'll start to prove the above statements, first using the definitions then we'll introduce two theorems that are useful to decide if a function is uniformly continuous or not.

Example 34 $f(x) = x$ is uniformly continuous on $(-\infty, \infty)$.

Proof   Let $\epsilon > 0$, and $\delta = \epsilon$. Then $|x - y| < \delta$ implies $|x - y| < \epsilon$ and $|f(x) - f(y)| < \epsilon$, so f is uniformly continuous.

Example 35   Let $b \in \mathbb{R}$, then $f(x) = x^2$ is uniformly continuous on $(-b, b)$.

Discussion   We want to use $|x - y| < \delta$ to conclude that

$$|x^2 - y^2| < \epsilon.$$

Manipulate to see

$$-\epsilon < (x - y)(x + y) < \epsilon,$$

and note that $x + y \le 2b$ so that

$$-\epsilon < \delta 2b < \epsilon,$$

and then $\delta = \frac{\epsilon}{|2b|}$.

Proof   Let $\epsilon > 0$, and $\delta = \frac{\epsilon}{|2b|}$. Then

$$|x - y| < \frac{\epsilon}{|2b|}$$

$$|2b||x - y| < \epsilon$$

$$|x + y||x - y| \leq |2b||x - y| < \epsilon$$

$$|x^2 - y^2| < \epsilon \quad \square$$

Example 36 $f(x) = x^2$ is not uniformly continuous on $(-\infty, \infty)$.

Discussion   Uniform continuity says

$$\forall \epsilon > 0 \, \exists \, \delta > 0 \text{ such that } \left(x, y \in (-\infty, \infty) \text{ and } |x - y| < \delta\right) \implies |x^2 - y^2| < \epsilon.$$

The negation is then

$$\exists \, \epsilon > 0 \, \forall \, \delta > 0 \text{ such that } \left(x, y \in (-\infty, \infty) \text{ and } |x - y| < \delta\right) \implies |x^2 - y^2| \geq \epsilon.$$

So we need to pick x, y so that $|x - y| < \delta$ will give us that $|x^2 - y^2| \geq \epsilon$. So we should make them some distance apart, but it must be less than δ. Lets try $\delta/2$, so that $y - x = \delta/2$. Then $y = x + \delta/2$ and

$$|x^2 - y^2| = \left|x^2 - \left(x + \frac{\delta}{2}\right)^2\right| = \left|x\delta + \frac{\delta^2}{4}\right|,$$

so that choosing $x = 1/\delta$ results in

$$\left|1 + \frac{\delta^2}{4}\right| \geq 1.$$

Proof   Let $\epsilon = 1$, $x = 1/\delta$, and $y = 1/\delta + \delta/2$. Then

$$|x - y| = |1/\delta - (1/\delta + \delta/2)| = \delta/2 < \delta,$$

and

$$|x^2 - y^2| = \left|\left(\frac{1}{\delta}\right)^2 - \left(\frac{1}{\delta} + \frac{\delta}{2}\right)^2\right| = \left|1 + \frac{\delta^2}{4}\right| \geq 1 \quad \square$$

Next we will relate Cauchy sequences to the concept of uniform continuity. Recall that the Cauchy sequence converges without us necessarily knowing what it converges to. This differs slightly from the usual convergence where we need to know the point of convergence to apply the definition. The theorem is relatively straightforward, but the application uses the contrapositive to conclude that a function is not uniformly continuous.

Uniformly continuous functions preserve Cauchy If f is uniformly continuous on D, and x_n is Cauchy in D, then $f(x_n)$ is Cauchy.

Proof Let x_n be a Cauchy sequence in D, and f uniformly continuous on D. Then for all $\delta > 0$ there exists $N \in \mathbb{N}$ such that

$$|x_m - x_n| < \delta.$$

Since f is uniformly continuous this implies

$$|f(x_m) - f(x_n)| < \epsilon$$

and $f(x_n)$ is Cauchy $\square$

The contrapositive statement is that if x_n is Cauchy and $f(x_n)$ is not Cauchy, then f is not uniformly continuous. Here is how we apply this:

Example 37 $f(x) = 1/x$ is not uniformly continuous on $(0, 1]$.

Proof   Let $x_n = 1/n$ and it is obviously Cauchy. Then

$$f(x_n) = \frac{1}{\frac{1}{n}} = n$$

which is not Cauchy. Therefore f is not uniformly continuous $\square$

Example 38 $f(x) = \sin(1/x)$ is not uniformly continuous on $(0, 1]$.

Proof Let $x_n = \frac{1}{n\pi}$. Then x_n is Cauchy but

$$\sin\left(\frac{1}{\frac{1}{n\pi}}\right) = \sin(n\pi) = 1, -1, 1, -1, 1, -1, \ldots$$

which is not Cauchy. Therefore $\sin(1/x)$ is not uniformly continuous on $(0, 1]$ □

Now we need to show that a continuous function on a closed and bounded interval is uniformly continuous. Then our other tool, the extension theorem, will follow.

Continuous on $[a, b]$ implies uniform continuity Let $f : [a, b] \to R$ be a continuous function. Then f is uniformly continuous.

Proof We'll argue by contradiction. Suppose f is not uniformly continuous. Then there exists $\epsilon > 0$ such that for all $\delta > 0$ we have

$$|x - y| < \delta \text{ and } |f(x) - f(y)| \geq \epsilon.$$

Let x_n, y_n be two convergent sequences in $[a, b]$. Then

$$|x_n - y_n| < \delta \text{ and } |f(x_n) - f(y_n)| \geq \epsilon.$$

Since $x_n \in [a, b]$, by Bolzano Weierstrass there is a convergent subsequence x_{n_k} converging to $x_0 \in [a, b]$ (since $[a, b]$ is closed). Since x_n is convergent, it also converges to x_0. Now for $\delta = 1/n$ we have

$$\frac{-1}{n} \leq y_n - x_n \leq \frac{1}{n},$$

$$\frac{-1}{n} + x_n \leq y_n \leq \frac{1}{n} + x_n,$$

and by the squeeze lemma $y_n \to x_0$. Then since f is continuous

$$(f(x_n) - f(y_n)) \to (f(x_0) - f(x_0)) = 0.$$

This contradicts $|f(x_n) - f(y_n)| \geq \epsilon > 0$ from above. Therefore if f is continuous on $[a, b]$ then f is uniformly continuous on $[a, b]$ □

The next property of uniformly continuous functions that we'll exploit has to do with the extension of a function. Basically we just add a point to the end of a function defined on an open interval so that it is defined on a closed and bounded interval. If we can do this, and the function remains continuous, then the image of the function will be closed and bounded. And it will turn out that the function is uniformly continuous.

Extension of a function Let $f : D \to R$ be a function. Then $\tilde{f}$ is an extension of f if the domain of f is a subset of the domain of $\tilde{f}$, and $x \in D \implies f(x) = \tilde{f}(x)$

Example 39 Let $f(x) = \sqrt{x}$ on $(0, 1)$. Then one extension is

$$\tilde{f}(x) = \begin{cases} 0 & x = 0 \\ \sqrt{x} & 0 < x < 1 \\ 1 & x = 1 \end{cases}$$

and another perfectly fine extension is

$$\tilde{f}(x) = \begin{cases} 2 & x = 0 \\ \sqrt{x} & 0 < x < 1 \\ 3 & x = 1 \end{cases}$$

Extension theorem Let $f : (a, b) \to R$ be continuous. Then f is uniformly continuous if and only if there exists an extension, $\tilde{f}$ on $[a, b]$ that is continuous on $[a, b]$.

Proof Suppose $\tilde{f}$ is continuous on $[a, b]$, then by the previous theorem, $\tilde{f}$ is uniformly continuous on $[a, b]$. Since $(a, b) \subset [a, b]$, and $f = \tilde{f}$ on (a, b), we have that f is uniformly continuous on (a, b).

Now suppose f is uniformly continuous on (a, b). Let x_n be a cauchy sequence in (a, b) converging to a. Since f is uniformly continuous, $f(x_n)$ is Cauchy, and therefore convergent. Define $\tilde{f}(a) = \lim f(x_n)$. Now for the other end of the interval do the same thing. Let y_n be a cauchy sequence in (a, b) converging to b. Since f is uniformly continuous, $f(y_n)$ is Cauchy, and therefore convergent. Define $\tilde{f}(b) = \lim f(y_n)$. Now $\tilde{f}$ is continuous on $[a, b]$ $\square$

Example 40 $f(x) = \sqrt{x}$ is uniformly continuous on $(0, 1)$.

Proof Extend $f(x)$ as follows:

$$\tilde{f}(x) = \begin{cases} 0 & x = 0 \\ \sqrt{x} & 0 < x < 1 \\ 1 & x = 1 \end{cases}$$

Then since $\tilde{f}$ is continuous on $[0, 1]$ it is uniformly continuous on $[0, 1]_\square$

In the next chapter we'll need uniform continuity to prove that continuous functions are integrable.

This page intentionally left blank.

Chapter 5

Integration, then Derivative

The underlying concept of an integral is area. This sounds simple enough, but when our area is defined by a real valued function considerable difficulty arises. As a result, several prominent mathematicians have developed alternatives. They include Riemann, Darboux, Stieltjes, LeBesgue, Daniell, and others. Considerable time could be spent studying the unique properties of each type separately. We will focus on the Darboux integral, with the assumption that the reader has some experience with Riemann sums. Recall the form of the Riemann sum:

$$R_n = \sum_{i=1}^{n} f(x_i^*)(x_i - x_{i-1}).$$

Where f is a function while $(x_i - x_{i-1})$ defines the width of the base of a rectangle. The point x_i^* is any point in the interval $[x_{i-1}, x_i]$. That x_i^* is *any* point in the interval makes things difficult to prove for each x_i^*. What Darboux does for us is to develop the idea in such a way that we only concern ourselves with the largest and smallest values of the function on each subinterval. In the graph on the next page the largest is depicted as the horizontal dotted line, and the smallest as the horizontal solid line (not the x-axis).

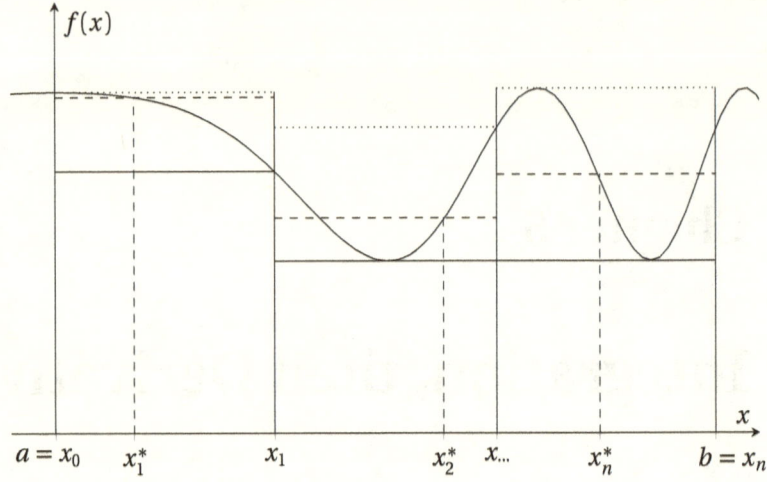

$$\cdots \quad M_i = \sup\{f(x) \mid x \in [x_{i-1}, x_i]\}$$
$$- \quad m_i = \inf\{f(x) \mid x \in [x_{i-1}, x_i]\}$$

Then to more precisely define the sub-intervals and allow varying widths we use the idea of a partition.

Partition   Let $a < b$ be real numbers and $a = x_0 < x_1 < x_2 < \cdots < x_n = b$. Then a partition is a finite ordered set

$$P = \{x_0, x_1, x_2, \ldots, x_n\}.$$

Now instead of the Riemann sum with the pesky x_i^* we have:

Darboux sums   For a function on a closed and bounded domain, $[a, b]$, and a partition, P, the upper and lower Darboux sums are:

$$U(f, P) = \sum_{i=1}^{n} M_i(x_i - x_{i-1}),$$

$$L(f, P) = \sum_{i=1}^{n} m_i(x_i - x_{i-1}).$$

Then by allowing various partitions of the interval we can define an upper and lower Darboux integral as the infimum or supremum as follows:

Upper and lower Darboux integral

$$U(f) = \inf \{U(f, P) \mid P \text{ is a partition of } [a, b]\}$$

$$L(f) = \sup \{L(f, P) \mid P \text{ is a partition of } [a, b]\}$$

Darboux integral When $U(f) = L(f)$ we say that the Darboux integral exists and use the usual symbol:

$$\int_a^b f(x)dx.$$

Note that by the completeness of the real numbers we can pick and choose partitions as we like. Additionally, whenever the function is defined, the upper and lower Darboux sums exist. Likewise, completeness ensures that the upper and lower Darboux integrals always exist as well. However, the Darboux integral may not exist since we don't know that $U(f)$ always equals $L(f)$. Consider the following example:

Dirichlet function The Dirichlet function on $[0, 1]$

$$f(x) = \begin{cases} 0 & x \notin \mathbb{Q} \\ 1 & x \in \mathbb{Q} \end{cases}$$

is not integrable. Every subinterval of any partition contains both rational and irrational numbers so that

$$U(f) = 1 \neq 0 = L(f).$$

This pathological example motivates the precise definition of a function given earlier and is, historically, the source of the varied approaches to integration. As mentioned earlier, much time could be spent exploring the details of various approaches to integration. The choice to present only the Darboux integral reflects my opinion that it best illustrates the challenge in defining what we mean when we talk about the area under a function between a and b. For a thorough discussion of the limitations of the Darboux and Riemann integrals, consult A Radical Approach to LeBesgue's Theory of Integration by David Bressoud. So in this chapter we will proceed with a little less discussion and weaker versions of some theorems than other books present. There's no point in spending too much time on the Darboux and Riemann theory that has been largely supplanted by LeBesgue, Daniell, and Haar. However we do need to establish the basic properties common to all integrals and so we proceed.

It would be nice if some of the properties of sums carried over to the integral, and indeed they do. In order to establish these properties we need to develop some precise machinery to work with.

Refinement   Let P and P_1 be partitions. If $x_i \in P$ implies $x_i \in P_1$, then P_1 is a refinement of P.

Intuitively we want to refine our partitions so that the number of rectangles gets larger and larger. A simple way to do this is to construct the sums with evenly spaced partitions and take a limit. The next example shows the details for the function $f(x) = x^3$ on a positive interval. Note that only in this example do we assume an evenly spaced partition.

Example 41   Find the upper and lower Darboux integrals for $f(x) = x^3$ on $[0, b]$ using an evenly spaced partition:

a solution   Recall the definition of the lower sum,

$$L(f, P) = \sum_{i=1}^{n} m_i(x_i - x_{i-1}).$$

Since f is increasing, m_i occurs at the left endpoint of the interval with value $f(x_{i-1})$. With n sub-intervals in $[0, b]$, if they are evenly spaced they have length $b/n = (x_i - x_{i-1})$. Then the left endpoint of each interval is $\frac{b}{n}(i-1)$. Applying these facts yields:

$$\sum_{i=1}^{n} f(x_{i-1})\frac{b}{n} = \sum_{i=1}^{n} \left(\frac{b(i-1)}{n}\right)^3 \frac{b}{n} = \frac{b^4}{n^4} \sum_{i=0}^{n-1} i^3.$$

Then from the known finite sum of cubes we have

$$\frac{b^4}{n^4}\sum_{i=0}^{n-1} i^3 = \frac{b^4}{n^4}(1+2+\cdots+n-1)^2,$$

and from the known finite sum of integers we have

$$\frac{b^4}{n^4}(1+2+\cdots+n-1)^2 = \frac{b^4}{n^4}\left(\frac{n(n-1)}{2}\right)^2 = \frac{b^4}{4n^4}(n^4-2n^3+n^2).$$

Take the limit as $n \to \infty$ and the lower integral is $b^4/4$.

Similarly the upper sum is computed as follows:

$$U(f,P)=\sum_{i=1}^{n} M_i(x_i - x_{i-1}).$$

Since f is increasing, M_i occurs at the right endpoint of the interval with value $f(x_i)$. With n sub-intervals in $[0,b]$, if they are evenly spaced they have length $b/n = (x_i - x_{i-1})$. Then the right endpoint of each interval is $\frac{bi}{n}$. Applying these facts yields:

$$\sum_{i=1}^{n} f(x_{i-1})\frac{b}{n} = \sum_{i=1}^{n}\left(\frac{bi}{n}\right)^3\frac{b}{n} = \frac{b^4}{n^4}\sum_{i=1}^{n} i^3.$$

Then from the known finite sum of cubes we have

$$\frac{b^4}{n^4}\sum_{i=}^{n} i^3 = \frac{b^4}{n^4}(1+2+\cdots+n)^2,$$

and from the known finite sum of integers we have

$$\frac{b^4}{n^4}(1+2+\cdots+n)^2 = \frac{b^4}{n^4}\left(\frac{n(n+1)}{2}\right)^2 = \frac{b^4}{4n^4}(n^4+2n^3+n^2).$$

Take the limit as $n \to \infty$ and the upper integral is $b^4/4$.

The next lemma establishes the relationship among upper and lower sums with different partitions.

Refinement lemma　Suppose $f : [a, b] \to \mathbb{R}$ is a bounded function so that there exists $m, M \in \mathbb{R}$ such that

$$m \leq f(x) \leq M \text{ for all } x \in [a, b].$$

Then for a partition P of $[a, b]$ we have

Part 1:　$m(b - a) \leq L(f, P) \leq U(f, P) \leq M(b - a).$

Part 2:　If P_1 is a refinement of P, then $L(f, P) \leq L(f, P_1) \leq U(f, P_1) \leq U(f, P).$

Part 3:　For any two partitions P, Q we have $L(f, P) \leq U(f, Q).$

Proof　Part 1: For the ith subinterval of $[a, b]$ from partition P, the basic definitions give us: $m \leq m_i \leq M_i \leq M$.
　　Then multiply by $(x_i - x_{i-1})$ and add up the terms

$$\sum_{i=1}^{n} m(x_i - x_{i-1}) \leq \sum_{i=1}^{n} m_i(x_i - x_{i-1}) \leq \sum_{i=1}^{n} M_i(x_i - x_{i-1}) \leq \sum_{i=1}^{n} M(x_i - x_{i-1}),$$

$$m(b - a) \leq L(f, P) \leq U(f, P) \leq M(b - a).$$

　　Part 2: Let P_1^k be the partition of $[x_{i-1}, x_i]$ induced by P_1. Then by part 1 we have

$$m_i(x_i - x_{i-1}) \leq L(f, P_1^k) \leq U(f, P_1^k) \leq M_i(x_i - x_{i-1}).$$

Now apply the induced partition on the ith subinterval, $[x_{i-1}, x_i]$ using the upper and lower sum definition. So let $x_k \in [x_{i-1}, x_i]$ and for the ith interval we have:

$$m_i(x_i - x_{i-1}) \leq \sum_{x_k \in (x_{i-1}, x_i]} m_k(x_k - x_{k-1}) \leq \sum_{x_k \in (x_{i-1}, x_i]} M_k(x_k - x_{k-1}) \leq M_i(x_i - x_{i-1}).$$

Now add up the sub intervals of $[a, b]$,

$$\sum_{i=1}^{n} m_i(x_i - x_{i-1}) \leq \sum_{i=1}^{n} \sum_{x_k \in (x_{i-1}, x_i]} m_k(x_k - x_{k-1}) \leq$$

$$\sum_{i=1}^{n} \sum_{x_k \in (x_{i-1}, x_i]} M_k(x_k - x_{k-1}) \leq \sum_{i=1}^{n} M_i(x_i - x_{i-1}).$$

Then apply the definitions of upper and lower sum to get

$$L(f,P) \le L(f,P_1) \le U(f,P_1) \le U(f,P).$$

Part 3: Let $W = P \cup Q$. Then part 2 gives us both

$$L(f,P) \le L(f,W) \le U(f,W) \le U(f,P),$$

$$L(f,Q) \le L(f,W) \le U(f,W) \le U(f,Q).$$

Therefore

$$L(f,Q) \le U(f,P) \; \square$$

The next theorem establishes upper and lower Darboux integrals in relation to upper and lower sums (and each other).

Bounds on Darboux integrals   Let $m, M \in \mathbb{R}$ such that $m \le f(x) \le M$ for all $x \in [a,b]$. Then

$$m(b-a) \le L(f) \le U(f) \le M(b-a).$$

Proof   Let P, Q be any partitions of $[a,b]$. Then by an application of the refinement lemma we have

$$m(b-a) \le L(f,P) \le U(f,Q) \le M(b-a).$$

Now take the infimum

$$\inf \{U(f,Q) \mid Q \text{ is a partition of } [a,b]\} = U(f).$$

Since $U(f)$ is the greatest lower bound we have that

$$m(b-a) \le L(f,P) \le U(f) \le U(f,Q) \le M(b-a).$$

Next, take the supremum

$$\sup \{L(f,P) \mid P \text{ is a partition of } [a,b]\} = L(f).$$

Since $L(f)$ is the least upper bound we have that

$$m(b-a) \le L(f,P) \le L(f) \le U(f) \le U(f,Q) \le M(b-a) \; \square$$

We're going to continue to use the infimum and supremum to establish the linearity of the integral. But first we have to show how the supremum and infimum are almost linear operators on sets. There is a trick to follow when multiplying a set by a negative number, as we'll see. It goes back to the proof of the corollary to the completeness axiom. It might be useful to review that proof at this time, before the next couple proofs. We will use the intuitive interpretation of a number times a set of real numbers as follows: $2*(0,3)=(0,6)$, or for another example

$$.5*\left\{\frac{1}{n} \mid n \in \mathbb{N}\right\} = \left\{\frac{1}{2n} \mid n \in \mathbb{N}\right\}.$$

5.1   Properties of the Integral

Constant multiple of supremum   Let S be a bounded non-empty set of real numbers and $k \geq 0$. Then

$$k \sup S = \sup kS$$

and

$$-k \sup S = \inf(-kS)$$

Discussion   The next several proofs continue to rely on the idea that the supremum is the *least* upper bound and that the infimum is the *greatest* lower bound. By combining this with some algebra manipulation we get the results we need.

Proof   For $k = 0$, the proof is trivial. So let $k > 0$ and by completeness both $\sup S$ and $\sup kS$ exist. By definition, for all $x \in S$ we have

$$x \leq \sup S,$$

$$kx \leq k \sup S.$$

Similarly with the set kS we have for all $kx \in kS$ that

$$kx \leq \sup kS.$$

Since $\sup kS$ is the *least* upper bound of kS we have that

$$\sup kS \leq k \sup S.$$

Now since $kx \leq \sup kS$, we have that

$$x \leq \frac{1}{k} \sup kS.$$

Since $\sup S$ is the *least* upper bound of S we have that

$$\sup S \leq \frac{1}{k} \sup kS.$$

Now

$$k \sup S \leq \sup kS.$$

Since we have shown non-strict inequality in both directions we conclude

$$k \sup S = \sup kS.$$

Next is the second equality. For all $x \in S$

$$\sup S \geq x,$$

$$-k \sup S \leq -kx.$$

Since $\inf(-kS)$ is the *greatest* lower bound we have

$$-k \sup S \leq \inf(-kS).$$

Now for all $-kx \in -kS$ we have

$$\inf(-kS) \leq -kx,$$

$$x \leq \frac{-1}{k} \inf(-kS).$$

Since $\sup S$ is the *least* upper bound

$$\sup S \leq \frac{-1}{k} \inf(-kS),$$

$$\inf(-kS) \leq -k \sup S.$$

Since we have shown non-strict inequality in both directions we conclude

$$\inf(-kS) = -k \sup S \; {}_\square$$

Naturally the above work will support the rule that the integral of a constant times a function is the constant times the integral of the function. The next theorem supports the conclusion that the integral of a sum of functions is the sum of the integral of the functions.

Supremum of sums   Let A, B be bounded non-empty subsets of real numbers. Then

$$\sup(A + B) = \sup A + \sup B.$$

Proof   Let $a \in A$ and $b \in B$; then $a + b \in A + B$. By definition of sup we have

$$a \leq \sup A$$

$$b \leq \sup B$$

Therefore

$$a + b \leq \sup A + \sup B.$$

And since $\sup(A + B)$ is the *least* upper bound of $a + b$ we have

$$\sup(A + B) \leq \sup A + \sup B.$$

To generate the opposite inequality, start with

$$a + b \leq \sup(A + B),$$

$$a \leq \sup(A + B) - b.$$

And since $\sup A$ is the *least* upper bound of a we have

$$a \leq \sup A \leq \sup(A + B) - b,$$

$$b \leq \sup(A + B) - \sup A.$$

And since $\sup B$ is the *least* upper bound of b we have

$$b \leq \sup B \leq \sup(A + B) - \sup A,$$

$$\sup B + \sup A \leq \sup(A + B).$$

Since we have shown non-strict inequality in both directions we conclude

$$\sup(A + B) = \sup A + \sup B \; _\square$$

A similar, omitted proof works for the statement concerning the infimum.

Infimum of sums Let A, B be bounded non-empty subsets of real numbers. Then

$$\inf(A + B) = \inf A + \inf B.$$

Next we show that upper and lower Darboux integrals are almost linear. It is almost because multiplying by a negative constant flips us between upper an lower integrals. Recall that we said the integral exists when both upper and lower have the same value. As a result, this flipping between upper and lower doesn't really matter when the integral exists.

Constant times integral Let $f : [a, b] \to \mathbb{R}$ be a bounded function, and $k \in \mathbb{R}$. Then when the integrals exist we have

$$k \int_a^b f(x)dx = \int_a^b kf(x)dx.$$

Proof Let P be any partition of $[a, b]$. Since the integral exists we have

$$L(kf) = \int_a^b kf(x)dx = U(kf),$$

$$\int_a^b kf(x)dx = \sup \{L(kf, P) \mid P \text{ is a partition of } [a, b]\}.$$

There are three cases (positive, negative, and zero) depending on k. We'll do the positive case first. Now focusing on the right hand side, by a basic property of sums we have

$$= \sup \{kL(f, P) \mid P \text{ is a partition of } [a, b]\}.$$

By constant multiple of a supremum

$$= k \sup \{L(f, P) \mid P \text{ is a partition of } [a, b]\}.$$

Since the integral exists we have

$$= kL(f) = k \int_a^b f(x)dx.$$

Now for negative k we start with

$$\int_a^b kf(x)dx = U(kf) = \inf\{U(kf,P) \mid P \text{ is a partition of } [a,b]\}.$$

Now focusing on the right hand side, by a basic property of sums we have

$$= \inf\{kU(f,P) \mid P \text{ is a partition of } [a,b]\}.$$

Then by constant multiple of a supremum for $k < 0$

$$= k\sup\{U(f,P) \mid P \text{ is a partition of } [a,b]\} = kU(f).$$

Since the integral exists we have

$$k\int_a^b f(x)dx = kU(f).$$

Since the case $k = 0$ is trivial, we are done $\square$

Integral of sums of functions   Let $f(x)$ and $g(x)$ be bounded functions on $[a,b]$. Then when the integrals exist

$$\int_a^b f(x)dx + \int_a^b g(x)dx = \int_a^b \big(f(x)+g(x)\big)dx.$$

Proof   Since the integral on the right exists we have

$$\int_a^b \big(f(x)+g(x)\big)dx = L(f+g) = \sup\{L(f+g,P) \mid P \text{ is a partition of } [a,b]\}.$$

By the property of sums and sets of real numbers

$$= \sup\{\{L(f,P)\} + \{L(g,P)\} \mid P \text{ is a partition of } [a,b]\}.$$

By supremum of sums

$$= \sup\{L(f,P) \mid P \text{ is a partition of } [a,b]\} + \sup\{L(g,P) \mid P \text{ is a partition of } [a,b]\}.$$

Then by the definition of Darboux sums and since the integrals exist

$$= L(f) + L(g) = \int_a^b f(x)dx + \int_a^b g(x)dx \ \square$$

Linearity of the integral   Let $f(x)$ and $g(x)$ be integrable on $[a,b]$. Then

$$k\int_a^b f(x)dx + k\int_a^b g(x)dx = \int_a^b k(f(x)+g(x))dx,$$

and the integral on the right exists.

Discussion   In each of the previous exercises, where we have demonstrated a property of integrals, we have assumed that the integral on the right side of the equation exists. However, it is not necessary to make that assumption. We need only assume that $\int_a^b f(x)dx$ and/or $\int_a^b g(x)dx$ exists. The proof becomes considerably more detailed to show that the resulting integral exists. Before we address the existence of an integral in some detail we'll prove a few more basic properties of the Darboux integral that should be familiar from calculus (and intuitive). Once we're finished with the basic properties, we'll justify the existence of the integral for monotonic and continuous functions.

Basic integral inequality   If $f(x) \le g(x)$ on $[a,b]$ and both are integrable, then

$$\int_a^b f(x)dx \le \int_a^b g(x)dx.$$

Proof   Let $h(x) = g(x) - f(x)$. Then $h(x) \ge 0$ and we have for all partitions P,

$$L(h,P) \ge 0.$$

Then as well for the supremum

$$L(h) \ge 0.$$

By definition of the Darboux integral and linearity we have

$$\int_a^b h(x)dx = \int_a^b g(x)dx - \int_a^b f(x)dx \ge 0,$$

$$\int_a^b g(x)dx \ge \int_a^b f(x)dx \quad \square$$

Integrals of combined domains   Suppose $\int_a^b f(x)dx$ exists and $a < c <$ b are all real numbers. Then if the integrals on the right exist

$$\int_a^b f(x)dx = \int_a^c f(x)dx + \int_c^b f(x)dx.$$

Proof   Consider the following partitions

$$P = \{a = x_0, x_1, \ldots x_n = b\},$$

$$P' = \{a = x_0, x_1, \ldots x_k, c, x_{k+1}, \ldots, x_n = b\} = P \cup \{c\},$$

$$Q = \{a = x_0, x_1, \ldots x_k, c\},$$

$$W = \{c, x_{k+1}, \ldots, x_n = b\}.$$

By the refinement lemma we have for any partition P,

$$L(f, P) \le L(f, P') \le U(f, P') \le U(f, P).$$

Since $P' = Q \cup W$ we have

$$L(f, P) \le L(f, Q \cup W) \le U(f, Q \cup W) \le U(f, P).$$

Since $Q \cap W = \{c\}$, and by summation properties, we have that

$$L(f, P) \le L(f, Q) + L(f, W) \le U(f, Q) + U(f, W) \le U(f, P).$$

Since our inequality holds for any partition P, and the partitions Q and W derive from P we can take the supremum of lower sums, and the infimum of upper sums. Since the integrals exist, we have

$$L(f) \le \int_a^c f(x)dx + \int_c^b f(x)dx \le U(f),$$

and

$$\int_a^b f(x)dx = \int_a^c f(x)dx + \int_c^b f(x)dx \;\square$$

In order to show that monotonic and continuous functions are integrable on a domain, $[a, b]$, we will need an ϵ criteria for integrability. This is developed next.

Criteria for integrability A bounded function, f on $[a, b]$ is integrable if and only if, for each $\epsilon > 0$ there exists a partition P such that

$$U(f, P) - L(f, P) < \epsilon.$$

Proof Suppose f is integrable. Let $\epsilon > 0$ be some small number. Then using completeness, there is a partition W such that

$$U(f, W) - \frac{\epsilon}{2} < U(f),$$

as well as a partition Q such that

$$L(F) < L(f, Q) + \frac{\epsilon}{2}.$$

Let $P = Q \cup W$ and apply the refinement lemma to get

$$U(f, P) - \frac{\epsilon}{2} \leq U(f, W) - \frac{\epsilon}{2} < U(f) = L(f) < L(f, Q) + \frac{\epsilon}{2} \leq L(f, P) + \frac{\epsilon}{2},$$

$$U(f, P) - \frac{\epsilon}{2} < L(f, P) + \frac{\epsilon}{2},$$

$$U(f, P) - L(f, P) < \epsilon.$$

Now, if for each $\epsilon >$ there is a partition P with

$$U(f, P) - L(f, P) < \epsilon.$$

We also have from the bounds on Darboux integrals

$$L(f, P) \leq L(f) \leq U(f) \leq U(f, P),$$

and so

$$U(f) - L(f) < \epsilon.$$

Since this holds for all $\epsilon > 0$, we have $U(f) = L(f)$ and f is integrable $\square$

For the next proof we'll need a minor new idea.

Mesh The mesh of a partition $P = \{a = x_0, x_1, \ldots, x_n = b\}$ is

$$\text{mesh}(P) = \sup\{(x_i - x_{i-1}) \mid x_i \in P\}.$$

Continuous functions are integrable   Let $f(x)$ be continuous on the closed and bounded interval $[a, b]$. Then $f(x)$ is integrable on $[a, b]$.

Proof   Since f is continuous on $[a, b]$, it is uniformly continuous. Let $x, y \in [a, b]$ and then for all $\epsilon > 0$ there exists a δ such that

$$|x - y| < \delta \implies |f(x) - f(y)| < \frac{\epsilon}{n(b - a)}.$$

Where n is the number of elements in P, such that $\text{mesh}(P) < \delta$. Also, by definition,

$$U(f, P) - L(f, P) = \sum_{i=1}^{n} M_i(x_i - x_{i-1}) - \sum_{i=1}^{n} m_i(x_i - x_{i-1}).$$

Since f is continuous, it takes its maximum and minimum values. So there exists $p_i \in [x_{i-1}, x_i]$ and $q_i \in [x_{i-1}, x_i]$ such that $M_i = f(p_i)$ and $m_i = f(q_i)$. We then have

$$U(f, P) - L(f, P) = \sum_{i=1}^{n} \big(f(p_i) - f(q_i) \big)(x_i - x_{i-1}).$$

Apply the bound given by uniform continuity to get

$$U(f, P) - L(f, P) < \frac{\epsilon}{(b - a)} \sum_{i=1}^{n} (x_i - x_{i-1}),$$

and then

$$U(f, P) - L(f, P) < \epsilon.$$

Therefore f is integrable. Since f was any continuous function, we conclude that continuous functions are integrable $\square$

Our geometric intuition tells us that since an integral can be interpreted as an area, that functions with some discontinuities ought to be integrable. For example, a well behaved step function ought to be integrable. In a popular presentation of LeBesgue integration, the step function is renamed a simple function, and used as a building block to define an integral that handles the Dirichlet function without any problems. For now we'll establish the integrability of monotonic functions (which could have as many discontinuities as you like).

Monotonic functions are integrable Let $f : [a, b] \to \mathbb{R}$ be monotonic. Then f is integrable.

Proof We'll just do the increasing, non-constant function case. Since f is monotonic we have for all $x \in [a, b]$ that $f(a) \le (f(x) \le f(b)$, and f is bounded. Let $\epsilon > 0$ and select a partition so that

$$\text{mesh}(P) < \frac{\epsilon}{f(b) - f(a)}.$$

Then applying the definitions of upper and lower sums yields

$$U(f, P) - L(f, P) = \sum_{i=1}^{n} M_i(x_i - x_{i-1}) - \sum_{i=1}^{n} m_i(x_i - x_{i-1}).$$

Since the function is monotonic and bounded, the supremum and infimum occur at the right and left endpoints of each interval. This yields

$$= \sum_{i=1}^{n} f(x_i)(x_i - x_{i-1}) - \sum_{i=1}^{n} f(x_{i-1})(x_i - x_{i-1}) = \sum_{i=1}^{n} \big(f(x_i) - f(x_{i-1})\big)(x_i - x_{i-1}).$$

Since $\text{mesh}(P) < \frac{\epsilon}{f(b) - f(a)}$ we have that

$$U(f, P) - L(f, P) < \sum_{i=1}^{n} \big(f(x_i) - f(x_{i-1})\big) \frac{\epsilon}{f(b) - f(a)},$$

and since the sum is telescoping

$$U(f, P) - L(f, P) < \big(f(b) - f(a)\big) \frac{\epsilon}{f(b) - f(a)} = \epsilon$$

So by the ϵ criteria for integration, f is integrable $\square$

5.2 Integration and Continuity

The reader that has some familiarity with the derivative might observe that the derivative of a function is not necessarily smoother than the function itself (think of sine and cosine, or e^x). Note also that it is fairly easy to come up with an example of a function that is not differentiable at a point, either by defining the function or by drawing it with a corner. In contrast, we really only have one example of a function that is not integrable in theory, and we can't effectively draw it. From a calculus course you probably have the impression that taking derivatives is easy and that computing integrals is harder. This is due to the types of functions and methods that you are taught (and are known). It turns out that integrating a function has a smoothing effect. To make this precise, we'll prove that the integral so far function is continuous, even if the function we are integrating is not continuous. As usual, we need to establish a preliminary result.

Integral absolute value inequality   When the integrals exist we have that

$$\left| \int_a^b f(x)dx \right| \le \int_a^b |f(x)| \, dx.$$

Proof   From the properties of absolute value we have

$$-|f(x)| \le f(x) \le |f(x)|.$$

Since the integrals exist, by the integral inequality (and linearity) we have

$$-\int_a^b |f(x)| \, dx \le \int_a^b f(x)dx \le \int_a^b |f(x)| \, dx.$$

But this is the definition of absolute value from example 20 so we have our result,

$$\left| \int_a^b f(x)dx \right| \le \int_a^b |f(x)| \, dx \; \square$$

Next we'll define the integral so far function, or you could call it the area so far function. Think back to adding up Riemann sums, starting from the left. We'll use the common notation of $F(x)$, and we will label it antiderivative, so you know where this is going. Be careful not to jump ahead though. We haven't defined a derivative yet, much less the limit of a function. Lets see what insights we get by taking this other approach.

Antiderivative   The antiderivative of an integrable function f on $[a, b]$ is

$$F(x) = \int_a^x f(t)dt,$$

defined for $a \leq x \leq b$. It may also be referred to as the area so far function, or the integral so far. Note the variable t is suppressed, it takes values between a and x.

Integrals are uniformly continuous   Let $f(x)$ be integrable on $[a, b]$. Then the following function is uniformly continuous on $[a, b]$.

$$F(x) = \int_a^x f(t)dt.$$

Proof   Let $M = \max\{\sup\{f(x)|x \in [a, b]\}, -\inf\{f(x)|x \in [a, b]\}\}$ and note that $f(x)$ is bounded by M. Also let $x < y$ both be in $[a, b]$ such that

$$|y - x| < \frac{\epsilon}{M}.$$

By algebra

$$M|y - x| < \epsilon.$$

Since $|f(t)| \leq M$

$$\int_x^y |f(t)|dt \leq M|y - x| < \epsilon.$$

By the absolute value integral inequality

$$\left|\int_x^y f(t)dt\right| \leq \int_x^y |f(t)|dt < \epsilon.$$

By an application of integrals of combined domains

$$|F(y) - F(x)| = \left|\int_x^y f(t)dt\right| < \epsilon.$$

Therefore $F(x)$ is uniformly continuous $\square$

Next we will prove the intermediate value theorem for integrals. It makes the assumption that the integrand, the function we are integrating, is continuous. Under this assumption we know that the integrand assumes the average value at least once. Naturally, we rely on the previous intermediate value theorem to make the conclusion.

Intermediate value theorem for integrals   Let f be continuous on $[a, b]$. Then there exists $c \in [a, b]$ such that

$$f(c)(b-a) = \int_a^b f(x) dx.$$

Proof   Since f is continuous on $[a, b]$, it is integrable on $[a, b]$, and attains its minimum, m, and maximum, M. From the bounds on darboux integrals we have

$$m(b-a) \le \int_a^b f(x) dx \le M(b-a).$$

By the extreme value theorem there exists y such that $f(y) = m$ and there exists z such that $f(z) = M$. Therefore

$$f(y)(b-a) \le \int_a^b f(x) dx \le f(z)(b-a),$$

$$f(y) \le \frac{1}{b-a} \int_a^b f(x) dx \le f(z).$$

By the intermediate value theorem there exists $c \in [a, b]$ such that

$$f(c) = \frac{1}{b-a} \int_a^b f(x) dx,$$

$$f(c)(b-a) = \int_a^b f(x) dx \quad \square$$

Let's return to the continuity of the integral. If the integral has this property that it smoothes out a function, that is, the integral of a discontinuous function may exist, and if it does exists, it must be continuous. Then it is natural to ask how smooth the resulting function may be. The mathematical concept that measures smoothness is the derivative. Indeed, when the derivative doesn't exist, the function is not smooth (there is a corner or discontinuity). We are now ahead of ourselves. Time to generate some more machinery.

5.3 Limits of Functions and the Derivative

The typical presentation of calculus starts with the idea of the limit of a function. However, as can be seen in this book so far, there is a lot of machinery that is masked over when one chooses to start with the limit of a function. In most calculus courses an $\epsilon - \delta$ definition of the limit is given, but the students largely rely on a sense of intuition about limits that is developed through the study of many examples. Here we continue to build from sequences of numbers as our conceptual building block.

Limit of a function   Let $f : D \to R$ be a real valued function. Let x_0 be a limit point of D, $L \in \mathbb{R}$, and $x_n \in D$. Then we write

$$\lim_{x \to x_0} f(x) = L$$

when for any convergent sequence $x_n \in D$ we have that $f(x_n) \to L$.

Since the limit of a function is defined in terms of sequences, the linear properties of sequences cary over directly, provided the limit exists in the first place. We'll simply state the linear property, but go ahead and prove the result for products and compositions of functions.

Limit of a function is linear   Suppose $\lim_{x \to x_0} f(x) = A$, $\lim_{x \to x_0} g(x) = B$, and $k, p \in \mathbb{R}$. Then

$$\lim_{x \to x_0} \left(k f(x) + p g(x) \right) = k A + p B.$$

If you have a pencil handy the proof fits in the margin below.

Products of limits   Suppose $\lim_{x \to x_0} f(x) = A$, and $\lim_{x \to x_0} g(x) = B$. Then

$$\lim_{x \to x_0} f(x)g(x) = AB.$$

Proof   Since the limits exist, for any sequence x_n converging to x_0 we have $f(x_n) \to A$, and $g(x_n) \to B$. Then by the product of limits of sequences property we have $f(x_n)g(x_n) \to AB$. Therefore

$$\lim_{x \to x_0} f(x)g(x) = AB \; \square$$

Composition of limits   Suppose that both

$$\lim_{x \to a} g(x) = g(a) = L, \text{ and } \lim_{x \to a} f(x) = L.$$

Then

$$\lim_{x \to a} g(f(x)) = g(L).$$

Proof   From the given limits, for $x_n \to a$ we have that $g(x_n) \to g(a)$, and so g is continuous. Since $f(x_n) \to L$, by the given continuity of g we have that $g(f(x_n)) \to g(L)$. Therefore

$$\lim_{x \to a} g(f(x)) = g(L) \; \square$$

In the next proof we'll need the idea of left and right handed limits, and the fact that a limit exists if and only if the left and right limits exist and are equal. Since we're assuming you have had a calculus course, we're going to omit the details here. However the outline would be to define the limit of a function from the left (or right) using sequences where $x_n \le x_0$ (or $x_n \ge x_0$). Then when both cases are true, any sequence is covered and we have the original meaning of the limit of a function. The right limit is decorated with a $+$ and the left with a $-$.

Derivative   The derivative of a function, $f : D \to R$, at $x_0 \in D$ is

$$f'(x_0) = \lim_{x \to x_0} \frac{f(x) - f(x_0)}{x - x_0}.$$

When the derivative exists at each point $x_o \in D$ we say the function is differentiable on D, and $f'(x)$ exists. Now that we have the precise machinery for the derivative, a measure of how smooth a function is, we can apply it to the integral so far function.

Fundamental theorem of calculus II   Suppose $f : [a, b] \to R$ is continuous. Let

$$F(x) = \int_a^x f(t)dt.$$

Then $F'(x) = f(x)$.

Proof   Since f is continuous, the integral exists, and $F(x)$ is continuous. Now to show the limit exists for an arbitrary $x_0 \in [a, b]$,

$$\lim_{x \to x_0} \frac{F(x) - F(x_0)}{x - x_0} = \lim_{x \to x_0} \frac{\int_a^x f(t)dt - \int_a^{x_0} f(t)dt}{x - x_0}.$$

Next we'll examine the right hand limit where $x \geq x_0$. Applying integrals of combined domains yields

$$\lim_{x \to x_0^+} \frac{\int_a^x f(t)dt - \int_a^{x_0} f(t)dt}{x - x_0} = \lim_{x \to x_0^+} \frac{1}{x - x_0} \int_{x_0}^x f(t)dt.$$

By the intermediate value theorem for integrals there exists $c \in [x_0, x]$ with

$$f(c) = \frac{1}{x - x_0} \int_{x_0}^x f(t)dt.$$

But since f is continuous, $\lim_{x \to x_0} f(x) = f(x_0) = f(c)$. Therefore

$$f(x_0) = \lim_{x \to x_0^+} \frac{1}{x - x_0} \int_{x_0}^x f(t)dt.$$

Combining the above result with the left hand limit case yields

$$f(x_0) = \lim_{x \to x_0} \frac{1}{x - x_0} \int_{x_0}^x f(t)dt.$$

Apply the definition of $F(x)$ to get

$$\lim_{x \to x_0} \frac{F(x) - F(x_0)}{x - x_0} = f(x_0) = F'(x_0),$$

and since x_0 is an arbitrary point in $[a, b]$ it holds for all $x \in [a, b]$ □

In words the previous theorem says that the derivative of the integral of a continuous function is the function. The next fundamental theorem of calculus says that the integral of the derivative of a function is the function. Note that in the previous theorem we used the intermediate value theorem for integrals, as a property of continuous functions. In order to prove the other part of the fundamental theorem of calculus, we need the mean value theorem which relies on Rolle's Theorem.

Rolle's theorem Suppose $f : (a, b) \to R$ is differentiable, continuous on $[a, b]$, and that $f(b) = f(a)$. Then there exists $c \in [a, b]$ such that $f'(c) = 0$.

Discussion This theorem at first appears to be narrowly focused, requiring the function to have the same elevation at the beginning and end of the interval. Then intuitively, the continuous function must have some point where it is horizontal so that it can return to the original elevation.

Proof Since f is continuous on $[a, b]$, by the extreme value theorem there exists $c \in [a, b]$ such that $f(c)$ is a maximum. Hence $f(c) \geq f(x)$ for all $x \in [a, b]$, and $f(c) - f(x) \geq 0$. Consider the right and left limits:

$$\lim_{x \to c^+} \frac{f(x) - f(c)}{x - c} \leq 0,$$

$$\lim_{x \to c^-} \frac{f(x) - f(c)}{x - c} \geq 0.$$

Since $f'(c)$ exists, the left and right limits must be $f'(c)$. Combining the above inequalities yields the result, $f'(c) = 0$ □

Note Since the basic rules of differentiation have short and straightforward proofs, we will not be including their proofs. Most calculus books cover their proofs, or have enough examples and exercises where the student uses the definition to compute a derivative. We'll resume a more detailed presentation in the next chapter with sequences of functions, and uniform convergence.

The strength of Rolle's theorem comes from the use of some creative algebra. We'll start with the equation of a line, modify it, and then subtract the original function in order to generate a new function that satisfies the requirements of Rolle's theorem. Consider the point-slope equation of the line parallel to the line between the points $(a, f(a)$ and $(b, f(b)$ where $y = g(x)$ and $y_0 = f(a)$:

$$g(x) = \frac{f(b) - f(a)}{b - a}(x - a) + f(a).$$

Then observe that $f(a) - g(a) = f(b) - g(b)$. We will apply Rolle's theorem to the function $f(x) - g(x)$ in order to prove the Mean Value Theorem.

Mean value theorem Suppose $f : (a, b) \to R$ is differentiable, and continuous on $[a, b]$. Then there exists $c \in [a, b]$ such that

$$f'(c) = \frac{f(b) - f(a)}{b - a}.$$

Proof Consider the following function

$$g(x) = \frac{f(b) - f(a)}{b - a}(x - a) + f(a).$$

It is continuous and differentiable because it is a line. By sums and constant multiples of continuous and differentiable functions the following function satisfies the requirements of Rolle's theorem

$$f(x) - g(x) = f(x) - \frac{f(b) - f(a)}{b - a}(x - a) - f(a).$$

Using a basic derivative rule at $x = c$ we have

$$f'(c) - g'(c) = f'(c) - \frac{f(b) - f(a)}{b - a} = 0,$$

$$f'(c) = \frac{f(b) - f(a)}{b - a} \quad \square$$

Notice how grouping some of the properties of the derivative with the proof of the fundamental theorem of calculus reinforces their relationship. We now have the tool to prove part one of the fundamental theorem of calculus. We won't have to assume that the derivative is continuous, as is done in several basic calculus books, where a shorter proof is given.

Fundamental theorem of calculus I   Suppose $f : [a, b] \to R$ is differentiable and f' is integrable. Then

$$\int_a^b f'(d)dx = f(b) - f(a).$$

Proof   Let $\epsilon > 0$, and since f' is integrable there exists a partition P such that

$$L(f', P) - U(f', P) < \epsilon.$$

Since f is differentiable, it is continuous (proof is below), and by the mean value theorem, for each sub interval there exists $c_i \in [x_{i-1}, x_i]$ such that

$$f(c_i) = \frac{f(x_i) - f(x_{i-1})}{x_i - x_{i-1}},$$

$$f(c_i)(x_i - x_{i-1}) = f(x_i) - f(x_{i-1}).$$

This quantity is bounded by upper and lower sums,

$$m_i(x_i - x_{i-1}) \leq f(c_i)(x_i - x_{i-1}) = f(x_i) - f(x_{i-1}) \leq M_i(x_i - x_{i-1}),$$

$$L(f', P) \leq \sum_{i=1}^{n} f(x_i) - f(x_{i-1}) \leq U(f', P).$$

Since the sum is telescoping, we have that

$$L(f', P) \leq f(b) - f(a) \leq U(f', P).$$

This inequality holds for all partitions so that

$$L(f') \leq f(b) - f(a) \leq U(f'),$$

and since the integral exists, $L(f') = U(f')$ and therefore

$$\int_a^b f'(d)dx = f(b) - f(a) \; \square$$

For the next proof we'll make a careful argument using limits. The novice student might make the mistake of assuming a limit exists, when that is not known.

Differentiable $\Longrightarrow$ continuous Let $f : D \to R$ be differentiable, then f is continuous.

Proof Since f is differentiable we have

$$f'(x) = \lim_{x \to x_0} \frac{f(x) - f(x_0)}{x - x_0}.$$

By basic limit laws we have

$$\lim_{x \to x_0} (x - x_0) = 0.$$

Since the limits exist, the product of the limits exist,

$$0 \cdot f'(x) = \lim_{x \to x_0} \frac{f(x) - f(x_0)}{x - x_0} \cdot (x - x_0),$$

$$0 = \lim_{x \to x_0} \left(f(x) - f(x_0) \right).$$

Since $f(x_0)$ is a constant,

$$f(x_0) = \lim_{x \to x_0} f(x).$$

Therefore f is continuous $\square$

L'Hospital's rule Suppose f and g are differentiable functions on an open interval such that $f(x_0) = 0$ and $g(x_0) = 0$. Then when the denominators are not zero

$$\lim_{x \to x_0} \frac{f(x)}{g(x)} = \frac{f'(x_0)}{g'(x_0)}.$$

Proof   Using the fact that $f(x_0) = 0$ and $g(x_0) = 0$, we have

$$\frac{f(x)}{g(x)} = \frac{f(x) - f(x_0)}{g(x) - g(x_0)} = \frac{\frac{f(x) - f(x_0)}{x - x_0}}{\frac{g(x) - g(x_0)}{x - x_0}}.$$

Then take the limit as x approaches x_0. Since the the limits on the right exist (they are the derivatives, $f'(x_0)$ and $g'(x_0)$).

$$\lim_{x \to x_0} \frac{f(x)}{g(x)} = \frac{\lim_{x \to x_0} \frac{f(x) - f(x_0)}{x - x_0}}{\lim_{x \to x_0} \frac{g(x) - g(x_0)}{x - x_0}} = \frac{f'(x_0)}{g'(x_0)} \quad \Box$$

A classic use of the derivative is to locate minimum or maximum values of a function. So first we need a precise definition of minimum and maximum.

Local maximum, minimum   Let $x_0 \in (a, b) \subset D$ with $f : D \to R$. If for all $x \in (a, b)$ we have that $f(x) \le f(x_0)$, then $f(x_0)$ is a local maximum at x_0. Likewise, if $f(x) \ge f(x_0)$ then $f(x_0)$ is a local minimum.

Min/Max $\Longrightarrow$ derivative is zero (when it exists)   Let $f : D \to R$ be a differentiable function with minimum or maximum at x_0. Then $f'(x_0) = 0$.

Proof   By definition

$$f'(x_0) = \lim_{x \to x_0} \frac{f(x) - f(x_0)}{x - x_0}.$$

If x_0 is a maximum, we have $f(x) - f(x_0) \le 0$. For $x \le x_0$ we then have

$$\lim_{x \to x_0^-} \frac{f(x) - f(x_0)}{x - x_0} \ge 0.$$

For $x \ge x_0$ we then have

$$\lim_{x \to x_0^+} \frac{f(x) - f(x_0)}{x - x_0} \le 0.$$

Since the function is differentiable, both limits exist and must be equal. Therefore $f'(x_0) = 0_\square$

Note that the converse statement is not true. For example $f(x) = x^3$ has $f'(0) = 0$, and $x = 0$ is not a local minimum or maximum. Another example to consider is a discontinuous function with a maximum as in

$$f(x) = \begin{cases} 1 & x = 0 \\ 0 & x \neq 0 \end{cases}$$

We also have continuous functions with a minimum where they are not differentiable as in $f(x) = |x|$. The point is that the restriction to differentiable functions in the above theorem is significant.

Next we will move on to some examples and counterexamples that highlight the interplay between the properties of the derivative, continuity, and extrema. Consider the function that was attempted to be plotted below.

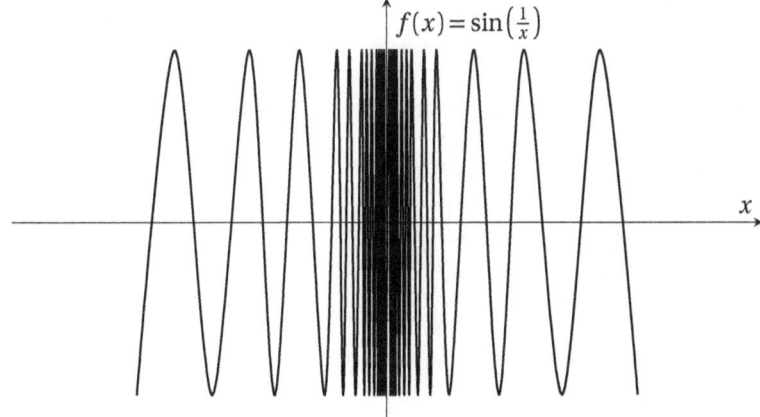

$$f(x) = \sin\left(\tfrac{1}{x}\right)$$

The astute observer will note that the function is not defined when $x = 0$. Is it possible to define the function value at $x = 0$ so that the function is continuous? Three options come to mind: 1, 0, or -1. Perhaps zero seems to be the natural choice, but what's wrong with 1 or -1?

Since continuity is defined in terms of sequences, we'll consider the sequences of zeroes, minimums, and maximums of our function. The zeroes occur at $x_n = \frac{1}{\pi n}$. To find the minimums and maximums, we could differentiate the expression and set the derivative to equal zero:

$$f'(x) = \frac{-1}{x^2} \cos\left(\frac{1}{x}\right) = 0,$$

$$-\cos\left(\frac{1}{x}\right) = 0.$$

Note that solving the above expression for x isn't a trivial exercise, especially since we haven't defined the cosine function yet. We could apply the secant method to find some roots, but we'll take a different approach. In the regular sine function the maximum occurs at $\frac{\pi}{2} + 2\pi n$, and the minimum at $\frac{3\pi}{2} + 2\pi n$. Since these values are half way between the zeroes, they are probably a good approximation for the the minimums and maximums of $\sin\left(\frac{1}{x}\right)$. That is, we guess that the maximums and minimums occur at

$$z_n = \frac{\frac{1}{n\pi} + \frac{1}{(n+1)\pi}}{2} = \frac{\pi(2n^2 + 2n)}{2n+1}.$$

This leads to the following table:

Iterate	z_n	$f(z_n) \approx$
1	$4\pi/3$	-0.866025
2	$12\pi/5$	0.951057
3	$24\pi/7$	-0.974928
4	$40\pi/9$	0.984808
5	$60\pi/11$	-0.989821
6	$84\pi/13$	0.992709
7	$112\pi/15$	-0.994522
8	$144\pi/17$	0.995734
9	$180\pi/19$	-0.996584
10	$220\pi/21$	0.997204

It appears that the odd iterations are converging to negative one and the even iterations are converging to positive one, that is

$$f(z_{\text{odd}}) \to -1 \text{ and } f(z_{\text{even}}) \to 1.$$

On the other hand z_n is diverging. This is not a problem, because $1/z_n$ converges to zero. Since the sine function is continuous we should have

$$\sin\left(\frac{1}{z_n}\right) \to 0.$$

But the table above contradicts this. No matter how we define f at zero, the function will be discontinuous. We know this because we have discovered two sequences, $1/z_{\text{odd}}$ and $1/z_{\text{even}}$ that both converge to zero, while the function of the sequences does not converge to the same point.

However, we can modify the expression slightly to get a continuous function, as follows:

$$g(x) = \begin{cases} x\sin\left(\frac{1}{x}\right) & x \neq 0 \\ 0 & x = 0 \end{cases}$$

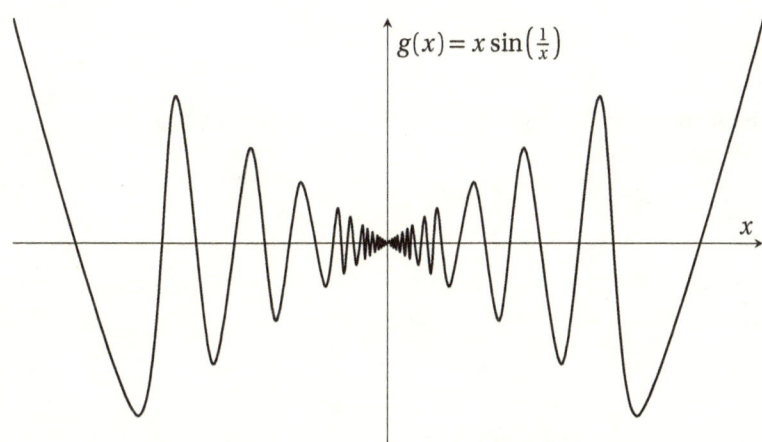

$$g(x) = x\sin\left(\frac{1}{x}\right)$$

In order to prove this function is continuous, we'll use a property of limits. Since limits depend on convergence, we give the proof for convergence and let the reader fill in the gap to draw the corresponding conclusion for limits of functions.

Bounded times zero limit Suppose y_n is bounded by M and $x_n \to 0$. Then $y_n x_n \to 0$.

Proof By convergence of x_n, for all $\epsilon/M > 0$ there exists $N \in \mathbb{N}$ such that $n > N$ implies

$$|x_n - 0| < \frac{\epsilon}{M}.$$

Since y_n is bounded we have

$$|y_n||x_n| < \frac{\epsilon}{M}M,$$

and then

$$|y_n x_n - 0| < \epsilon.$$

So that $y_n x_n \to 0$ $\square$

The similar statement for limits of functions is below (the proof is an application of the above result, and so is omitted).

Bounded times zero limit If $|g(x)| \le M$ and $\lim_{x \to x_0} f(x) = 0$, then

$$\lim_{x \to x_0} f(x)g(x) = 0.$$

Example 42 The following function is continuous at $x = 0$

$$g(x) = \begin{cases} x \sin\left(\frac{1}{x}\right) & x \ne 0 \\ 0 & x = 0 \end{cases}$$

Proof Let $x_n \to 0$. We know that $\left|\sin\left(\frac{1}{x_n}\right)\right| \le 1$. By the theorem above, the product of the sequences converges to zero,

$$x_n \sin\left(\frac{1}{x_n}\right) \to 0.$$

Since $g(0) = 0$, the function is continuous at $x = 0$ $\square$

To see that $g'(0)$ does not exist we show that the following limit does not exist

$$\lim_{x \to 0} \frac{g(x) - g(0)}{x - 0} = \lim_{x \to 0} \frac{x \sin\left(\frac{1}{x}\right)}{x} = \lim_{x \to 0} \sin\left(\frac{1}{x}\right).$$

The limit on the right was shown not to exist a couple pages ago. Therefore the derivative does not exist.

We started with $\sin(1/x)$ and showed that it was not continuous, no matter how we define it at $x = 0$. By modifying it slightly we see how to get a continuous function that is not differentiable. We'll take one more step to get the function to be differentiable. Consider

$$h(x) = \begin{cases} x^2 \sin\left(\frac{1}{x}\right) & x \neq 0 \\ 0 & x = 0 \end{cases}$$

It is differentiable at $x = 0$ because

$$\lim_{x \to 0} \frac{h(x) - h(0)}{x - 0} = \lim_{x \to 0} \frac{x^2 \sin\left(\frac{1}{x}\right)}{x} = \lim_{x \to 0} x \sin\left(\frac{1}{x}\right) = 0.$$

So $h'(x)$ exists, however the second derivative, $h''(x)$ is not differentiable at the origin because a form similar to $g(x)$ above is the result.

The next theorem is the chain rule. Due to the varied expressions we have for composition of functions, learning to apply the chain rule gets all the attention in calculus class. For example the following expression is the composition of four functions, but the functions are represented in four different ways

$$f(x) = \sqrt{(3x - 4)^3}.$$

If you go back to your calculus book to look at a proof, some magical transformation is pulled out of a hat to make the proof work. The below presentation follows the article written by Stephen Kenton in the College Mathematics Journal.

Chain rule   Let $f : (a, b) \to (c, d)$ and $g : (c, d) \to R$. Suppose f is differentiable at $z \in (a, b)$ and g is differentiable at $f(z)$. Then

$$\left(g(f(x)\right)'(z) = g'(f(z))f'(z).$$

Proof Outline   Apply the definition and some algebra to get

$$\left(g(f(x)\right)'(z) = \lim_{x \to z} \frac{g(f(x)) - g(f(z))}{x - z} = \lim_{x \to z} \frac{g(f(x)) - g(f(z))}{f(x) - f(z)} \frac{f(x) - f(z)}{x - z}.$$

Since f is differentiable at z, the below limit exists

$$\lim_{x \to z} \frac{f(x) - f(z)}{x - z}.$$

Since g is differentiable at $f(z)$, when $f(x)-f(z)\neq 0$ the below limit exists

$$\lim_{x\to z}\frac{g(f(x))-g(f(z))}{f(x)-f(z)}=g'(f(z)).$$

By the product of limits rule, we have the result when there is no zero in the denominator. Now we consider $f(x)-f(z)=0$. In this case $g'(f(z))$ still exists, so it has some finite value, that is, it is bounded. But then also we have

$$\lim_{x\to z}\frac{f(x)-f(z)}{x-z}=\lim_{x\to z}\frac{0}{x-z}=0.$$

In this case, we have reduced the limit definition of the derivative we started with, to a product of a limit that is bounded with a limit approaching zero. Hence the result is zero and the limit exists $\square$

5.4   The Sine Function as a Motivating Example

Back in section 1.4 we defined a polynomial. A slight modification will give us the definition of a power series.

Power series   A power series $p(x):\mathbb{R}\to\mathbb{R}$ about the point a has the following form:

$$p(x)=\sum_{n=0}^{\infty}c_n(x-a)^n.$$

Where x,a,c_n are all real numbers and n is an integer. In layman's terms the power series is a polynomial with infinite powers.

Back in trigonometry or algebra II class we learned that the sine function is the ratio of the opposite side to the hypotenuse of a right triangle. This turned out to be useful, but it isn't very precise. Then in calculus we learned the properties of the derivative of $f(x)=\sin(x)$:

$$f'(x)=\cos(x),$$

$$f''(x)=-\sin(x),$$

$$f'''(x)=-\cos(x),$$

$$f^{(4)}(x) = \sin(x).$$

No one bothered to rigorously justify these claims, but we believed them based on pictures and what our calculators told us. Here we will begin the work to justify these claims, and at the same time develop a rigorous definition for the sine function. These techniques reveal what your calculator uses to compute values for the trigonometric, exponential, and logarithmic functions. In addition they can be used to justify and/or develop trigonometric identities. The basic idea starts with an obvious equation,

$$\sin(x) - \sin(x) = 0.$$

Then we apply the above derivative properties of $f(x) = \sin(x)$, and $f''(x)$ to get a differential equation:

$$f''(x) + f(x) = 0.$$

Recall from calculus that a differential equation is an equation where the function is the unknown, and at least one derivative of the function is in the equation. The problem is to find the function that has the relationship specified by the differential equation. But don't we know that $f(x) = \sin(x)$ is the solution for the differential equation we just wrote? Perhaps...but we don't really know what $\sin(x)$ is. How is it computed? What is the rule that we apply to x to get the value in the range?

We will use a common method for solving differential equations, guess and check. Our guess will be that the solution is a power series about $a = 0$, and we will use the properties of power series that we have not yet justified.

- For some values of x the power series converges to a number, making it reasonable to treat a power series as a function.

- The derivative of a power series can be computed by differentiating the terms in the series.

- two power series are equal if the coefficients, c_n are equal when the exponents of the terms are equal.

Now we'll start computing. Our guess for a solution is

$$f(x) = \sum_{n=0}^{\infty} c_n x^n.$$

Then the derivatives are

$$f'(x) = \sum_{n=0}^{\infty} n c_n x^{n-1},$$

$$f''(x) = \sum_{n=0}^{\infty} n(n-1) c_n x^{n-2}.$$

And our differential equation says $f''(x) + f(x) = 0$ or

$$\sum_{n=0}^{\infty} n(n-1) c_n x^{n-2} + \sum_{n=0}^{\infty} c_n x^n = 0.$$

We need to re-index our sums so that the exponents of x are the same. One way to do this is below:

$$\sum_{n=0}^{\infty} (n+2)(n+1) c_{n+2} x^n + \sum_{n=0}^{\infty} c_n x^n = 0.$$

Then we factor out the x^n term because we want to solve for the c_n, as they are what really defines the particular power series:

$$\sum_{n=0}^{\infty} [(n+2)(n+1) c_{n+2} + c_n] x^n = 0.$$

Then we must have

$$(n+2)(n+1) c_{n+2} + c_n = 0,$$

$$c_{n+2} = \frac{-c_n}{(n+2)(n+1)}.$$

The coefficients are then

$$c_2 = \frac{-c_0}{2 \cdot 1}$$

$$c_3 = \frac{-c_1}{3 \cdot 2}$$

$$c_4 = \frac{-c_2}{4 \cdot 3} = \frac{c_0}{4!}$$

$$c_5 = \frac{-c_3}{5 \cdot 4} = \frac{c_1}{5!}$$

$$c_6 = \frac{-c_4}{6 \cdot 5} = \frac{-c_0}{6!}$$

$$\vdots$$

But since we are interested in the sine function which we know has $\sin(0) = 0$, our power series must have this same property. That is

$$f(0) = \sum_{n=0}^{\infty} c_n 0^n = c_0 = 0.$$

Then the only terms remaining in the series solution are those with odd index. The solution is

$$f(x) = \sum_{n=0}^{\infty} \frac{(-1)^n c_1}{(2n+1)!} x^{2n+1}.$$

Applying a second initial condition, $f'(0) = \cos(0) = 1$ yields (from $0^0 = 1$)

$$f'(0) = \sum_{n=0}^{\infty} \frac{(-1)^n c_1 (2n+1)}{(2n+1)!} 0^{2n} = c_1 = 1.$$

Then we finally have a power series definition for the sine function.

$$\sin(x) = \sum_{n=0}^{\infty} \frac{(-1)^n}{(2n+1)!} x^{2n+1}.$$

When our calculator computes $\sin(1$ radian$)$ it could use the first several terms of the above series. By hand the calculations are

$$\sin(1) \approx 1 - \tfrac{1}{3!} + \tfrac{1}{5!} - \tfrac{1}{7!} + \tfrac{1}{9!}$$

$$\approx 1 - 0.166\bar{6} + 0.008\bar{3}3 - 0.00019841\ldots + 0.0000027557\ldots$$

$$\approx 0.84148\ldots$$

$$\sin(1) = 0.8414709848\ldots$$

Note that the convergence is quite rapid, that the fifth term we are adding has five leading zeros in the decimal, thanks to the factorial in the denominator. This same process can be used to define the cosine function

by applying different initial conditions on the differential equation. For
the exponential function the differential equation is

$$f'(x) = f(x) \text{ with initial condition } f(0) = 1.$$

In the next chapter we will resume a more rigorous approach than we
have had in the current section. We'll also use more examples to answer
three basic questions about the continuity, differentiability, and integrability of sequences of functions. This example with the sine function is
just one special case where the properties of the power series (which is
just a special type of sequence of functions) turn out to be very nice. We
will see that this is often not the case in practical examples.

Chapter 6

Sequences of Functions

There are three questions that are the theme of this chapter and they all have to do with sequences of functions, $f_n(x)$, or a series

$$\sum_{n=0}^{\infty} f_n(x).$$

Note that a series is just a sequence of partial sums.

Q1: If all the $f_n(x)$ are continuous, and $f_n(x) \to f(x)$, is $f(x)$ necessarily continuous?

Q2: If all the $f_n(x)$ are differentiable, and $f_n(x) \to f(x)$, is $f(x)$ necessarily differentiable? Furthermore, is the limit of the derivatives the derivative of the limits? Written another way, is the following true?

$$\lim_{n \to \infty} \frac{d}{dx} f_n(x) = \frac{d}{dx} \lim_{n \to \infty} f_n(x)$$

Q3: If all the $f_n(x)$ are integrable, and $f_n(x) \to f(x)$, is $f(x)$ necessarily integrable? Furthermore, is the limit of the integrals the integral of the limits? Written another way, is the following true?

$$\lim_{n \to \infty} \int_a^b f_n(x) dx = \int_a^b \lim_{n \to \infty} f_n(x) dx$$

Here we'll assume you know some of the basic limits of the exponential and some rational functions.

6.1   Pointwise Convergence

Here are some examples that illustrate the limitations of pointwise convergence.

Counterexample Q1   We note that $f_n(x) = x^n$ is a continuous function on $[0,1]$. For $x = 1$ we have

$$\lim_{n \to \infty} 1^n = 1.$$

However for any other value of $x \in [0,1)$

$$\lim_{n \to \infty} x^n = 0.$$

So each $f_n(x)$ is continuous because each one is a polynomial. However the sequence of functions converges to

$$f(x) = \begin{cases} 0 & x \in [0,1) \\ 1 & x = 1 \end{cases}$$

which is not continuous. Hence question one is answered in the negative.

Counterexample Q2   Consider $f_n(x) = e^{-nx^2}$ on $[-1,1]$. Each function in the sequence is differentiable. However when we examine limits we find that for $x = 0$,

$$\lim_{n \to \infty} e^{-n0^2} = e^0 = 1.$$

And for $x \neq 0$ we have

$$\lim_{n \to \infty} e^{-nx^2} = \lim_{n \to \infty} \frac{1}{e^{nx^2}} = 0.$$

And even though each function in the sequence of functions is differentiable, the sequence converges to

$$f(x) = \begin{cases} 0 & x \in [-1,0) \cup (0,1] \\ 1 & x = 0 \end{cases}$$

Therefore the answer to question two is negative. The limit of the sequence is not even continuous, so we can't even consider whether the limit of the derivatives is the derivative of the limits.

Counterexample Q3 Consider $f_n(x) = nxe^{-nx^2}$ by examining the limit when $x \neq 0$

$$\lim_{n \to \infty} nxe^{-nx^2} = \lim_{n \to \infty} \frac{nx}{e^{nx^2}}.$$

Apply L'Hospital's rule (differentiating with respect to n) to get

$$\lim_{n \to \infty} \frac{x}{x^2 e^{nx^2}} = \lim_{n \to \infty} \frac{1}{xe^{nx^2}} = 0.$$

Note that $f_n(0) = 0$ so that $f_n \to 0$. Then

$$\int_0^1 \lim_{n \to \infty} nxe^{-nx^2} dx = \int_0^1 0 dx = 0.$$

However, for

$$\lim_{n \to \infty} \int_0^1 nxe^{-nx^2} dx$$

we must integrate first. A u-substitution with $u = -nx^2$ yields

$$\lim_{n \to \infty} \int_0^1 \frac{-1}{2} e^u du = \lim_{n \to \infty} \frac{-1}{2} e^u \bigg|_0^1 = \lim_{n \to \infty} \frac{-1}{2} e^{-nx^2} \bigg|_0^1 = \lim_{n \to \infty} \frac{1-e^{-n}}{2} = \frac{1}{2}.$$

So in this example we found that the limit of the sequence of functions is integrable. However the integral of the limits is not the limit of the integrals.

These last three examples illustrate the limitations of the pointwise convergence we have been using.

Pointwise convergence The sequence $f_n(x)$ converges pointwise to $f(x)$ if for each x in the domain and for each $\epsilon > 0$ there exists $N \in \mathbb{N}$ such that $n > N$ implies

$$|f_n(x) - f(x)| < \epsilon.$$

Here N depends on x and ϵ. A stronger type of convergence is uniform convergence where N only depends on ϵ. Note the analogous situation with continuity and uniform continuity.

6.2   Uniform Convergence

Uniform convergence   Let $f_n(x): D \to R$ be a sequence of functions. If for each $\epsilon > 0$ there exists $N \in \mathbb{N}$ such that $n > N$ implies

$$\left|f_n(x) - f(x)\right| < \epsilon$$

for all $x \in D$, then we say $f_n(x)$ converges uniformly to $f(x)$, or $f_n(x) \to f(x)$ uniformly.

To see that a sequence of functions does not converge uniformly, we look for values of x that depend on n in a way that causes

$$\left|f_n(x) - f(x)\right| > 0.$$

Counterexample Q1 is not uniformly convergent   For $f_n(x) = x^n$, recall that it converges to

$$f(x) = \begin{cases} 0 & x \in [0, 1) \\ 1 & x = 1 \end{cases}$$

Then for $x \neq 1$ we want

$$x^n - 0 > 0,$$

or even better if

$$x^n > \frac{1}{4},$$

$$x > \left(\frac{1}{4}\right)^{\frac{1}{n}}.$$

So choose $x_n = \left(\frac{1}{4}\right)^{\frac{1}{n}}$ and we have that $x_n \neq 1$ and

$$f_n\left(\left(\frac{1}{4}\right)^{\frac{1}{n}}\right) = \left(\frac{1}{4}\right).$$

We have determined x_n in D such that

$$\left|f_n(x_n) - f(x_n)\right| = \left|\frac{1}{4} - 0\right| > 0.$$

Therefore $f_n(x)$ is not uniformly convergent.

Counterexample Q2 is not uniformly convergent   Recall that $f_n(x) = e^{-nx^2}$ on $[-1, 1]$ converged pointwise to zero except at $x = 0$. So we'll look for a sequence, x_n converging to zero such that $f_n(x_n) \neq 0$. That is, solve for x in terms of n so that

$$e^{-nx^2} - 0 > 0.$$

If we solve explicitly we'll end up with a logarithm, but that isn't necessary. Just look at nx^2. If $x_n = \frac{1}{\sqrt{n}}$, then $x_n \to 0$, and $nx_n^2 = 1$. So that

$$f_n(x_n) = e^{-n\left(\frac{1}{\sqrt{n}}\right)^2} = \frac{1}{e} > 0.$$

Therefore $f_n(x)$ is not uniformly convergent.

Counterexample Q3 is not uniformly convergent   Recall that $f_n(x) = nxe^{-nx^2}$ converges pointwise to $f(x) = 0$. Since $f(x)$ is continuous, there isn't an obvious point to go after for determining a sequence, x_n, to make $f_n(x) > 0$. We'll use a different approach by finding the maximum values of $f_n(x)$. Taking the derivative with respect to x, and equating it to zero hopefully yields an extreme value

$$f_n'(x) = nxe^{-nx^2}(-2nx) + ne^{-nx^2} = 0,$$

$$e^{-nx^2}\left(-2n^2x^2 + n\right) = 0,$$

$$2n^2x^2 = n$$

$$x_n = \frac{1}{\sqrt{2n}} \to 0.$$

Now $f_n(x_n)$ is as follows

$$\frac{n}{\sqrt{2n}}e^{-n\left(\frac{1}{\sqrt{2n}}\right)^2} = \frac{n}{\sqrt{2n}}e^{\frac{-1}{2}},$$

and is unbounded. For $n \geq 2$ we have

$$\left|f_n(x_n) - f(x)\right| = \left|\frac{n}{\sqrt{2n}}e^{\frac{-1}{2}} - 0\right| \geq \frac{1}{\sqrt{e}} > 0.$$

Therefore $f_n(x)$ is not uniformly convergent. Notice that the integral of $f(x)$ exists, but that the limit of the integrals is not the integral of the limit. We will have an example later where the limit of the integrals is the integral of the limit. But first we'll establish the usefulness of uniform convergence.

Uniform convergence preserves continuity   Suppose $f_n(x): D \to R$ are continuous functions, uniformly convergent to $f(x)$. Then $f(x)$ is continuous.

Discussion   The insight comes from writing the last part of the definitions, first for the definition for continuity of the individual functions in the sequence,

$$\left|f_n(x) - f_n(x_0)\right| < \frac{\epsilon}{3}.$$

Next for the definition for uniform convergence at both x and x_0

$$\left|f_n(x) - f(x)\right| < \frac{\epsilon}{3}, \text{ and } \left|f_n(x_0) - f(x_0)\right| < \frac{\epsilon}{3}.$$

And finally, for the desired result

$$\left|f(x) - f(x_0)\right| < \epsilon.$$

Then apply the triangle inequality to the three former inequalities to get the latter one.

Proof   Let $\epsilon > 0$. From the given convergence we have that for $n > N$ and from the given continuity that for $|x - x_0| < \delta$

$$\left|f(x) - f(x_0)\right| \le \left|f(x) - f_n(x) + f_n(x) - f_n(x_0) + f_n(x_0) - f(x_0)\right|$$

$$\le \left|f(x) - f_n(x)\right| + \left|f_n(x) - f_n(x_0)\right| + \left|f_n(x_0) - f(x_0)\right| < \frac{\epsilon}{3} + \frac{\epsilon}{3} + \frac{\epsilon}{3} = \epsilon.$$

Therefore $f(x)$ is continuous at x_0. Since x_0 is an arbitrary point in D, $f(x)$ is continuous on D $\square$

A very similar argument works to show that uniform convergence preserves uniform continuity as well. Next we'll prove the theorem that justifies the technique used in the examples to show that they are not uniformly convergent. Recall that we found values of x such that

$$\left|f_n(x) - f(x)\right| > 0.$$

This can be precisely stated as follows:

Supremum criteria for uniform convergence $f_n(x) \to f(x)$ uniformly on D if and only if

$$\lim_{n \to \infty} \sup\{|f_n(x) - f(x)| \mid x \in D\} = 0.$$

Proof   Let $\epsilon = 1/N$ and then for sufficiently large $n > N$

$$|f_n(x) - f(x)| < 1/N$$

for all $x \in D$. Therefore

$$0 \le \sup\{|f_n(x) - f(x)| \mid x \in D\} < 1/N.$$

Since this is true for all $\epsilon > 0$ we can let $N \to \infty$. And since $n > N$ we have

$$0 \le \lim_{n \to \infty} \sup\{|f_n(x) - f(x)| \mid x \in D\} \le \lim_{N \to \infty} 1/N = 0.$$

By the squeeze lemma

$$\lim_{n \to \infty} \sup\{|f_n(x) - f(x)| \mid x \in D\} = 0.$$

Now for the other direction the existence of the limit means that for any $\epsilon > 0$ there is an N such that $n > N$ implies

$$|\sup\{|f_n(x) - f(x)| \mid x \in D\} - 0| < \epsilon,$$

so that N does not depend on x, and we have for all $x \in D$

$$|f_n(x) - f(x)| < \epsilon.$$

Therefore $f_n(x) \to f(x)$ uniformly $\square$

Now whenever $f_n(x) - f(x)$ is differentiable, we can use a derivative to find good guesses for minimum or maximum points as we did in the examples. We have shown several examples without uniform convergence, now for a couple examples with uniform convergence (and the proof). In the first one we will rely on the definition of uniform convergence because $f(x)$ won't be differentiable. For the following one we'll formally use the above theorem.

Example 43   Let $f_n(x) = \sqrt{x^2 + 1/n}$ for $x \in \mathbb{R}$. Prove $f_n(x)$ converges uniformly.

Discussion   The first step is to make a good guess about what the function will converge to. A crucial insight is that $1/n \to 0$, and so we intuitively ignore that part (since it isn't in a denominator, where it would require attention). What remains is $\sqrt{x^2}$ which almost equals x, except that it is always positive. The other function with this property is $f(x) = |x|$, so it ought to be what we're converging to. Finding a computer to graph $f_1(x), f_2(x), f_3(x)$, and $f_4(x)$ reinforces our suspicions.

The scratch work starts with the conclusion we want, working backwards, then seeing that the convergence of the sequence $1/n$ directly relates to the convergence of the sequence of functions $f_n(x)$, and is independent of x.

Proof   It should be obvious that

$$\sqrt{x^2} - |x| = 0.$$

Since $1/n \to 0$ and the positive square root function is continuous

$$\lim_{n \to \infty} \sqrt{x^2 + \frac{1}{n}} - |x| = 0.$$

This means for all $\epsilon > 0$ there exists $N \in \mathbb{N}$ such that $n > N$ implies

$$\left| \sqrt{x^2 + 1/n} - |x| - 0 \right| < \epsilon.$$

Note that N does not depend on x in any way and

$$\left| \sqrt{x^2 + 1/n} - |x| \right| < \epsilon.$$

Therefore $f_n(x) \to f(x)$ uniformly $\square$

For the next example we'll apply the previous theorem.

Example 44 Let $f_n(x) = \frac{x}{1+nx^2}$ for $x \in \mathbb{R}$. Prove that $f_n(x)$ is uniformly convergent.

Discussion To decide what this will converge to, we note that the only n is in the denominator so that

$$\lim_{n \to \infty} \frac{x}{1+nx^2} = 0.$$

Proof Since $f_n(x)$ is differentiable we'll take the derivative (note that only the numerator is needed) to find minimum or maximum.

$$f_n'(x) = 1 + nx^2 - x(2nx) = 1 - nx^2 = 0,$$

$$1 = nx^2 \implies x = \pm\frac{1}{\sqrt{n}}.$$

Then evaluating

$$\lim_{n \to \infty} \sup\left\{\left|\frac{x}{1+nx^2} - 0\right| \, x \in D\right\}$$

at the critical points yields

$$\lim_{n \to \infty} \sup\left\{\left|\frac{\pm\frac{1}{\sqrt{n}}}{1+n\left(\pm\frac{1}{\sqrt{n}}\right)^2} - 0\right| \, x \in D\right\},$$

$$\lim_{n \to \infty} \sup\left\{\left|\frac{1}{2\sqrt{n}}\right| \, x \in D\right\} = 0.$$

Therefore $f_n(x) \to 0$ □

At the beginning of section 5.2 we mentioned that integrating a function has a smoothing effect, and then shortly after that we proved that the integral so far of a function, when the integral exists, is continuous. We also gave example 42, an apparently smooth function whose derivative does not exist. These results indicate that proving that the integral of the limit is the limit of the integral will be easier than the similar property for the derivative. Indeed this is the case, and our proof will follow the theme of the proof that uniform convergence preserves continuity.

Uniform convergence preserves integrability   Let $f_n(x) \to f(x)$ uniformly, and suppose that the $f_n(x)$ are each integrable. Then $f(x)$ is integrable.

Proof   First we'll use uniform convergence to establish a needed inequality, and then use it in an epsilon over three argument. Let $\epsilon/3 > 0$ and since $f_n(x)$ is integrable there are partitions, P_n of $[a, b]$ where $i = 0, 1, 2, \ldots k$ indexes the elements of the partition with the following supremums and infimums:

$$M_{in} = \sup\{f_n(x) \mid x \in [x_{i-1}, x_i]\},$$

$$m_{in} = \inf\{f_n(x) \mid x \in [x_{i-1}, x_i]\}.$$

By uniform convergence there exists $N \in \mathbb{N}$ such that $n > N$ implies that for all $x \in [a, b]$

$$|f_n(x) - f(x)| < \epsilon,$$

$$-\epsilon < f_n(x) - f(x) < \epsilon,$$

$$-\epsilon + f(x) < f_n(x) < \epsilon + f(x).$$

For the partition

$$P = \bigcup_{n=0}^{N} P_n,$$

since M_{in} is the least upper bound we have

$$-\epsilon + f(x) < f_n(x) \le M_{in} < \epsilon + f(x),$$

$$f(x) < f_n(x) + \epsilon \le M_{in} + \epsilon < 2\epsilon + f(x).$$

And since M_i is the least upper bound of $f(x)$ we have

$$f(x) \le M_i < f_n(x) + \epsilon \le M_{in} + \epsilon < 2c + f(x).$$

Next subtract ϵ and $f(x)$ to get

$$-\epsilon \le M_i - \epsilon - f(x) < f_n(x) - f(x) \le M_{in} - f(x) < \epsilon.$$

A careful application of inequalities yields

$$0 \le (M_{in} - f(x)) - (M_i - \epsilon - f(x)) \le 2\epsilon.$$

$$-\epsilon \leq M_{in} - M_i \leq \epsilon.$$

$$|M_{in} - M_i| \leq \epsilon.$$

Then a very similar argument applies to the infimum yielding

$$|m_{in} - m_i| \leq \epsilon.$$

Now we move to the $\epsilon/3$ argument. By integrability we have

$$\left| U(f_n, P) - L(f_n, P) \right| < \epsilon/3.$$

Next, consider

$$\left| L(f_n, P) - L(f, P) \right| = \left| \sum_{i=0}^{k} m_{in}(x_i - x_{i-1}) - \sum_{i=0}^{k} m_i(x_i - x_{i-1}) \right|,$$

$$= \left| \sum_{i=0}^{k} m_{in}(x_i - x_{i-1}) - m_i(x_i - x_{i-1}) \right| = \left| \sum_{i=0}^{k} (m_{in} - m_i)(x_i - x_{i-1}) \right|.$$

By the triangle inequality we have

$$\leq \sum_{i=0}^{k} |(m_{in} - m_i)| |(x_i - x_{i-1})|.$$

From the top of the page we have $|m_{in} - m_i| \leq \epsilon/3|b - a|$,

$$\leq \sum_{i=0}^{k} \frac{\epsilon}{3|b - a|} |(x_i - x_{i-1})| = \epsilon/3.$$

For the upper sums, a similar argument to the one above yields

$$\left| U(f_n, P) - U(f, P) \right| \leq \epsilon/3.$$

Applying the triangle inequality to the three established inequalities yields

$$\left| U(f, P) - L(f, P) \right| = \left| U(f_n, P) - L(f_n, P) + L(f_n, P) - L(f, P) + U(f, P) - U(f_n, P) \right|$$

$$\leq \left| U(f_n, P) - L(f_n, P) \right| + \left| L(f_n, P) - L(f, P) \right| + \left| U(f, P) - U(f_n, P) \right|$$

$$< \epsilon/3 + \epsilon/3 + \epsilon/3 = \epsilon.$$

We have met the criteria for integrability, and therefore $f(x)$ is integrable □

Uniform convergence allows interchange of limit and integral   Let $f_n(x) \to$ $f(x)$ uniformly, and suppose that the $f_n(x)$ are each integrable. Then

$$\lim_{n \to \infty} \int_a^b f_n(x) dx = \int_a^b f(x) dx$$

Proof   The previous theorem established the existence of the integral on the right. To see that the limit of the integrals is the integral of the limits we apply the criteria for uniform convergence with the absolute value integral inequality and a basic property of integrals:

$$\left| \int_a^b f(x) dx - \int_a^b f_n(x) dx \right| \le \left| \int_a^b f(x) - f_n(x) dx \right| \le \int_a^b |f(x) - f_n(x)| dx$$

$$\le \lim_{n \to \infty} \sup \{ |f_n(x) - f(x)| \ | x \in D \} (b - a) = 0 \, \square$$

The previous theorem gives us a sufficient condition to conclude that we can interchange limits with integrals, but it is not necessary. Consider the next example of a sequence of functions that are not uniformly convergent. However, with this particular function we can interchange limit with integral and get the same result.

Example 45   Let $g_n(x) = \frac{nx}{1+n^2 x^2}$ Then since there is an n^2 in the denominator but only a lower power of n in the numerator we have point wise convergence, that is

$$\lim_{n \to \infty} g_n(x) = 0.$$

Since the functions are all differentiable, we take a derivative to find critical points.

$$g_n'(x) = \frac{n - n^3 x^2}{(1 + n^2 x^2)^2} = 0$$

yields $x = \frac{1}{n}$ as a critical point, so that

$$g_n \left(\frac{1}{n} \right) = \frac{1}{2}$$

is a maximum. Hence $g_n(x)$ is not uniformly convergent to 0. However if we compute the limit of the integral we use a substitution $u = 1 + n^2 x^2$, and

$$\lim_{n\to\infty} \int_0^1 \frac{nx}{1+n^2x^2}dx = \lim_{n\to\infty} \frac{\ln(1+n^2x^2)}{2n}\bigg|_0^1 = \lim_{n\to\infty} \frac{\ln(1+n^2)}{2n} = 0.$$

This example indicates to us that the previous theorem may not be very strong. By not strong we mean that there are many cases where the conclusion of the theorem is true but the requirements of the theorem are not met. A stronger version is presented next.

Dominated convergence theorem Let $f_n(x) \to f(x)$ pointwise, and suppose the f_n are all integrable. If there is a function $g(x)$ such that $f_n(x) \le g(x)$ for all n and for all x, then $f(x)$ is integrable and

$$\int f(x)d\mu = \lim_{n\to\infty} \int f_n(x)d\mu.$$

We will not prove the dominated convergence theorem because it is not true for Darboux or Riemann integrals. This is symbolized by the use of $d\mu$ instead of dx in the above theorem. An example illustrating this fact uses r_n, an enumeration of the rational numbers such as the Calkin-Wilf formula given on page ##.

$$f_n(x) = \begin{cases} 1 & x \in \{r_1, r_2, \ldots, r_n, r_{n+1}, \ldots\} \subset \mathbb{Q} \\ 0 & x \notin \mathbb{Q} \end{cases}$$

Then the limit of $f_n(x)$ is the dirichlet function, shown to not be Darboux integrable in chapter 5. However $f_n(x)$ meets the criteria of the dominated convergence theorem. The dominated convergence theorem is true for LeBesgue integrals which rely on measure theory, and so the above statement isn't entirely precise. Don't worry, when both the Lebesgue and Darboux integrals exist, they have the same value.

The Riemann or Darboux approach to integration was developed from Cauchy's finite sum approach (from the Greek method of exhaustion) where the area under a curve was approximated by left endpoints as in

$$\text{Area} \approx \sum_{i=1}^n f(x_i)(x_{i+1} - x_i).$$

So the essence was to take a horizontal segment of length $(x_{i+1} - x_i)$, and then ask how high is the function above the segment? The initial answer was $f(x_i)$. This requires one answer for the height, so that the next step of multiplication can be done to get an area above the segment. The error is that only constant functions have one height. Riemann's insight was to allow any $x_i^* \in [x_i, x_{i+1}]$. But this still assumes that a constant function is a good enough approximation (will have a vanishing error in the limit) and it is in many cases, and all was good until Dirichlet's function was taken seriously (and other problems with Fourier series arose).

What Lebesgue does is recognize that more attention needs to be paid to the height of the function. He asks, how big is the set of x–values for which $y_i \le f(x) \le y_{i+1}$? This can be written as

$$\mu(A_i) = \left| \{x \mid y_i \le f(x) \le y_{i+1}\} \right|.$$

Then the Lebesgue approximating sum is

$$\sum_{i=1}^{n} y_i \mu(A_i).$$

The work to precisely define the size of a set of points on the real line, or in $\mathbb{R}^n$, is called measure theory. The first, probably third of a course in real analysis will be devoted to measure theory, and it requires a robust understanding of set theory. Perhaps you're next book or math course will be on real analysis, the euphemism for an introduction to measure theory and Lebesgue integration.

To summarize the answer to question three, observe that example 41 is not uniformly convergent, but it is pointwise convergent and it is bounded. The dominated convergence theorem tells us that we can interchange limits and integrals, as we discovered by computing them. On the next page will be the last example for limits and integration. It turns out to be pointwise convergent, but not uniformly convergent or bounded. Even though this dominated convergence theorem does not apply, the limit and integral will interchange. The dominated convergence theorem is a sufficient, but not necessary condition. It turns out to be a much more useful criteria than uniform convergence. There is a necessary and sufficient condition called the Vitali convergence theorem, but its statement uses the language of measure theory and so it is omitted.

Example 46 For $h(x) = 0$ and

$$h_n(x) = \frac{n^2 x}{1 + n^3 x^2},$$

we see that pointwise convergence is to zero (divide top an bottom by n^2). So we have

$$\int_0^1 \lim_{n \to \infty} h_n(x) = 0.$$

On the other hand, using the substitution $u = 1 + n^3 x^2$ yields

$$\lim_{n \to \infty} \int_0^1 \frac{n^2 x}{1 + n^3 x^2} dx = \lim_{n \to \infty} \frac{\ln(1 + n^3 x^2)}{2n} \Big|_0^1 = \lim_{n \to \infty} \frac{\ln(1 + n^3)}{2n} = 0.$$

If the dominated convergence theorem were a necessary condition, we would not be able to show that $h_n(x)$ is not bounded everywhere, as we are about to do,

$$h_n'(x) = n^2 + n^5 x^2 - 2n^5 x^2 = n^2 - n^5 x^2 = 0,$$

$$x_n = n^{\frac{-3}{2}}.$$

Then evaluating our function at this critical point and taking the limit yields

$$\lim_{n \to \infty} h_n\left(n^{\frac{-3}{2}}\right) = \lim_{n \to \infty} \frac{n^2 n^{\frac{-3}{2}}}{1 + n^3 n^{-3}} = \lim_{n \to \infty} \frac{\sqrt{n}}{2} = \infty.$$

What measure theory does is introduce the concept called, almost everywhere. This provides the necessary nuance to ignore the subsets of the domain where a function is not well behaved, when the measure of that subset is zero.

Finally, we will answer question two, about the interchange of limits and derivatives. The key insight for this proof is to abandon the definition of the derivative, and use the fundamental theorem of calculus. The resulting proof is quite slick.

Interchange of limit and derivative   Let $f_n(x)$ be a sequence of differentiable functions with continuous derivatives, pointwise convergent to $f(x)$. If $f'_n(x) \to g(x)$ uniformly, then $g(x) = f'(x)$ and $f'(x)$ is continuous.

Proof   Fix x_0 in the domain. Since $f'_n(x)$ is continuous, it is integrable and by the fundamental theorem of calculus,

$$\int_{x_0}^{x} f'_n(t) dt = f_n(x) - f_n(x_0).$$

Since f'_n converges uniformly to g, taking the limit as $n \to \infty$ yields

$$\int_{x_0}^{x} g(t) dt = f(x) - f(x_0),$$

$$f(x) = f(x_0) + \int_{x_0}^{x} g(t) dt.$$

Taking the derivative yields

$$f'(x) = g(x) \,_\square$$

Example 47   Let $f_n(x) = \frac{x}{1+n^2 x^2}$ for $x \in \mathbb{R}$. We have pointwise convergence of the functions,

$$\lim_{n \to \infty} \frac{x}{1+n^2 x^2} = 0.$$

They have continuous derivatives but the limit of the derivatives is

$$\lim_{n \to \infty} f'_n(x) = \lim_{n \to \infty} \frac{1 - n^2 x^2}{(1 + n^2 x^2)^2} = 0,$$

except at $x = 0$ where $f'_n(0) = 1$. So $f'(x)$ is not continuous. We can conclude that $f'_n(x)$ does not converge uniformly.

The final example for this section will be one that illustrates that our condition for interchange of limit and derivative is not sufficient.

Example 48   Consider the following sequence of functions with $f_n(0) = 0$ and for $x \neq 0$

$$f_n(x) = \frac{1}{n}\left(x^2 \sin\left(\frac{1}{x}\right)\right).$$

Based on previous work, each $f_n(x)$ is differentiable. However

$$f_n'(x) = \frac{1}{n}\left(2x \sin\left(\frac{1}{x}\right) - \cos\left(\frac{1}{x}\right)\right)$$

is not continuous. However taking limits yields

$$\lim_{n\to\infty} f_n(x) = 0,$$

$$\lim_{n\to\infty} f_n'(x) = 0,$$

so that the limit of the derivatives is the derivative of the limits.

6.3 Series

In this section we establish some facts that will be used in the last two sections of the book on power series and Taylor series. Often these convergence results are examples or exercises given separately from where they are used in the general theory. The reader could skip this section and come back to it as needed when reading about power series and Taylor series.

Harmonic Series Diverges   Let $a_n = 1/n$ then

$$\sum_{n=1}^{\infty} \frac{1}{n} = \infty.$$

There are many approaches available for this proof. Several rely on integration, and several rely on an intuitive inductive argument. This proof by contradiction is nice and short.

Proof   Suppose the harmonic series converges. Then let

$$H = \left(1 + \frac{1}{2}\right) + \left(\frac{1}{3} + \frac{1}{4}\right) + \left(\frac{1}{5} + \frac{1}{6}\right) \cdots$$

However

$$H = \left(1 + \frac{1}{2}\right) + \left(\frac{1}{3} + \frac{1}{4}\right) + \left(\frac{1}{5} + \frac{1}{6}\right) \cdots > \left(\frac{1}{2} + \frac{1}{2}\right) + \left(\frac{1}{4} + \frac{1}{4}\right) + \left(\frac{1}{6} + \frac{1}{6}\right) \cdots$$

$$= 1 + \frac{1}{2} + \frac{1}{3} + \frac{1}{4} + \cdots = H.$$

So if the harmonic series converges, we have $H > H$, which is clearly absurd. We conclude that the harmonic series diverges □

However the alternating harmonic does converge to $\ln 2$, that is

$$\sum_{n=1}^{\infty} \frac{(-1)^{n-1}}{n} = \ln 2.$$

This can be seen by computing the Taylor series for $\ln(1 + x)$. The Taylor series are introduced in section 6.5.

p-test for Convergence   If $p > 1$, then the following series converges:

$$\sum_{n=1}^{\infty} \frac{1}{n^p}.$$

A common (more intuitive) proof of this result uses integration, but it isn't necessary. However, the insight to factor a two from the denominator in order to bound the partial sums is a technique that was also used in the Heine-Borel theorem.

Proof   We will bound the partial sums, and then since they are increasing, we have a convergent sequence. Write the partial sums as

$$s_n = \sum_{k=1}^{n} \frac{1}{k^p} \leq 1 + \sum_{k=1}^{n} \frac{1}{(2k)^p} + \frac{1}{(2k+1)^p}.$$

Observe that

$$\frac{1}{(2k)^p} + \frac{1}{(2k+1)^p} < \frac{2}{(2k)^p},$$

and so

$$s_n < 1 + 2^{1-p} \sum_{k=1}^{n} \frac{1}{k^p} = 1 + 2^{1-p} s_n,$$

$$s_n < 1 + 2^{1-p} s_n,$$

$$s_n(1 - 2^{1-p}) < 1,$$

$$s_n < \frac{1}{1 - 2^{1-p}}.$$

Therefore s_n is bounded. Since all the terms in the series are positive, s_n is increasing. Since bounded, increasing sequences converge, s_n converges □

On the next page we will state and prove Raabe's test, which is slightly stronger than the ratio test which follows. The theme with these tests is the harmonic series, that it appears to be very close to the boundary between convergent and divergent series.

Raabe's Test   Let $a_n \neq 0$ be the terms of a series, and $c > 1$. Suppose that for $n \geq N$

$$\frac{|a_n|}{|a_{n-1}|} \leq 1 - \frac{c}{n}.$$

Then $\sum_{n=0}^{\infty} |a_n|$ converges. On the other hand, if $c \leq 1$ and

$$\frac{|a_n|}{|a_{n-1}|} \geq 1 - \frac{c}{n},$$

then $\sum_{n=0}^{\infty} |a_n|$ diverges.

Discussion   The insight needed for the convergence part of this proof is that if $n|a_n|$ is decreasing, then $N|a_N|$ is an upper bound not only for the terms with $n \geq N$ in the sequence, but for the series $\sum |a_n|$ as well. This idea drives our algebra to establish $n|a_n|$ is decreasing. It's lower bound should be obvious, zero. For the divergence part, we bound the series below by the harmonic series.

Proof   For convergence, start with the given inequality:

$$\frac{|a_n|}{|a_{n-1}|} \leq 1 - \frac{c}{n},$$

$$n|a_n| \leq (n-c)|a_{n-1}| = (n-1-(c-1))|a_{n-1}|,$$

$$n|a_n| \leq (n-1)|a_{n-1}| - (c-1)|a_{n-1}|.$$

So $n|a_n|$ is decreasing for $n \geq N$. Since $n|a_n|$ is always positive, it is bounded below by zero, and therefore convergent to some $M \in \mathbb{R}$. Next we make a telescoping series as follows:

$$(n-1)|a_{n-1}| - n|a_n| \geq (c-1)|a_{n-1}| \geq 0.$$

Observe the cancellation that happens when these two inequalities are added for the N and $N+1$ terms:

$$(N-1)|a_{N-1}| - N|a_N| \geq (c-1)|a_{N-1}| \geq 0,$$

$$(N)|a_N| - (N+1)|a_{N+1}| \geq (c-1)|a_N| \geq 0.$$

For a finite sum we add the first $N-1$ terms separately to get

$$\sum_{k=0}^{N-1}|a_k|-(n+1)|a_{n+1}|\geq(c-1)\sum_{k=N}^{n}|a_k|+\sum_{k=0}^{N-1}|a_k|\geq0.$$

Now take the limit as n approaches infinity, and N remains fixed

$$\sum_{k=0}^{N-1}|a_k|-\lim_{n\to\infty}n|a_n|\geq(c-1)\sum_{k=N}^{\infty}|a_k|+\sum_{k=0}^{N-1}|a_k|\geq0.$$

The left hand side is a finite sum plus the limit of a convergent sequence, bounding the series $\sum|a_n|$ with zero. Hence $\sum|a_n|$ converges. Now for the divergence conclusion we start with

$$\frac{|a_n|}{|a_{n-1}|}\geq1-\frac{c}{n},$$

$$n|a_n|\geq(n-c)|a_{n-1}|.$$

For sufficiently large n we have

$$n|a_n|\geq(n-c)|a_{n-1}|\geq(n-1)|a_{n-1}|,$$

so that $n|a_n|$ is increasing. Then there is some b such that

$$n|a_n|\geq b>0,$$

$$|a_n|\geq\frac{b}{n}.$$

Then add up the terms to get

$$\sum_{n=0}^{\infty}|a_n|\geq b\sum_{n=0}^{\infty}\frac{1}{n},$$

which diverges because the harmonic series diverges □

Raabe's test is a stronger and more flexible version of the ratio test that comes next. This means that we have to find the appropriate value of c in order to apply the test to a particular series. A natural choice is $c=1$, and the resulting, easier to apply test is the ratio test.

Ratio Test　　For a series, $\sum a_n$, if

$$\lim_{n \to \infty} \frac{|a_n|}{|a_{n-1}|} < 1,$$

then $\sum |a_n|$ converges. If the limit above is greater than or equal to one, then $\sum |a_n|$ diverges.

Proof　　From Raabe's test we have for $c > 1$ and $n \geq N$ that if

$$\frac{|a_n|}{|a_{n-1}|} \leq 1 - \frac{c}{n}.$$

Then $\sum_{n=0}^{\infty} |a_n|$ converges. We just move the strict inequality from a restriction on c to the central inequality in the limit, and let $c = 1$ so that if

$$\lim_{n \to \infty} \frac{|a_n|}{|a_{n-1}|} < \lim_{n \to \infty} 1 - \frac{1}{n} = 1,$$

then $\sum_{n=0}^{\infty} |a_n|$ converges. Similarly, from the other inequality in Raabe's test we get the divergence criteria □

In both of the convergence tests presented here, the conclusion is about the convergence of the sum of the absolute value of the terms. We call this absolute convergence. It is possible to converge without absolute convergence, when the signs of the terms are alternating positive and negative. When this is the case there is a phenomena where the series can sum to different values depending on the order of the terms in the series. This is not very intuitive, but it does happen. We give an example to motivate the detailed proof about double series that follows.

Rearrangements of series can give different results　　Consider the alternating harmonic series for $\ln 2$

$$\ln 2 = 1 - \frac{1}{2} + \frac{1}{3} - \frac{1}{4} + \frac{1}{5} - \frac{1}{6} + \frac{1}{7} - \frac{1}{8} \cdots$$

We can write 1/2 times the above convergent series as

$$\frac{1}{2} \ln 2 = 0 + \frac{1}{2} + 0 - \frac{1}{4} + 0 + \frac{1}{6} + 0 - \frac{1}{8}$$

Now if we add these two convergent series we get

$$\frac{3}{2}\ln 2 = 1 + 0 + \frac{1}{3} - \frac{1}{2} + \frac{1}{5} + 0 + \frac{1}{7} + \cdots$$

which is a rearrangement of the original series

$$1 + \frac{1}{3} - \frac{1}{2} + \frac{1}{5} + \frac{1}{7} - \frac{1}{4} + \frac{1}{9} + \frac{1}{11} - \frac{1}{6} \cdots$$

Now we will prove that the product of absolutely convergent series is absolutely convergent.

Multiplication of absolutely convergent series is stable   Suppose $\sum |a_n|$ and $\sum |b_m|$ converge. Then $\sum |a_n b_m|$ converges as well.

Proof   The given convergence can be stated as

$$A = \sup \left\{ \sum_{n=0}^{N} |a_n| \;\middle|\; N \in \mathbb{N} \right\} = \sum_{n=0}^{\infty} |a_n| > 0,$$

$$B = \sup \left\{ \sum_{m=0}^{M} |b_m| \;\middle|\; M \in \mathbb{N} \right\} = \sum_{m=0}^{\infty} |b_m| > 0.$$

Then by constant multiple of supremum, since $A > 0$ we have

$$AB = \sup \left\{ \sum_{m=0}^{M} A|b_m| \;\middle|\; M \in \mathbb{N} \right\}.$$

Since absolute convergence ensures each partial sum is positive, we also have (from the constant multiple of supremum again)

$$AB = \sup \left\{ \sum_{m=0}^{M} \sum_{n=0}^{N} |a_n||b_m| \;\middle|\; M, N \in \mathbb{N} \right\}.$$

Since the sums in the set are finite, associativity of addition yields

$$AB = \sup \left\{ \sum_{n=0}^{N} \sum_{m=0}^{M} |a_n||b_m| \;\middle|\; M, N \in \mathbb{N} \right\}.$$

Since the order of summation doesn't matter for the finite sums, we'll write this more compactly as

$$AB = \sup \left\{ \sum_{n=0}^{N} \sum_{m=0}^{M} |a_n||b_m| \;\middle|\; M, N \in \mathbb{N} \right\} = \sum_{m+n=0}^{\infty} |a_n||b_m| \; \square$$

There are two convergent sequences that we'll need to establish properties for power series, and here they are. They could have just as easily been presented back in chapter two.

Convergence of $n^{1/n}$ The sequence $n^{1/n}$ converges to 1.

Discussion   Recall the Binomial Theorem which states

$$(a+b)^n = a^n + na^{n-1}b + \frac{n(n-1)}{2}a^{n-2}b^2 + \cdots + nab^{n-1} + b^n.$$

The insight is that when a and b are both positive we can use the first several terms of the binomial expansion to establish an upper bound on b^2, and thus on b. Using a calculator, we guess that $n^{1/n}$ converges to one. The second insight is to define a and b so that a goes away, and b is in terms of the convergent sequence. This is not something that a novice would generate on the fly.

Proof   Let $s_n = n^{1/n} - 1$, and then $n = (1 + s_n)^n$, but by the binomial theorem we have

$$n = (1 + s_n)^n \geq 1 + n s_n + \frac{n(n-1)}{2} s_n^2 + \cdots,$$

$$n > \frac{n(n-1)}{2} s_n^2,$$

$$\frac{2n}{n(n-1)} > s_n^2,$$

$$\sqrt{\frac{2}{(n-1)}} > s_n.$$

Since the square root function is continuous, taking the limit as n approaches ∞ yields

$$0 \geq \lim_{n \to \infty} s_n.$$

However we also know that

$$n \geq 1,$$
$$n^{1/n} \geq 1,$$
$$n^{1/n} - 1 = s_n \geq 0.$$

So by the squeeze lemma, $s_n \to 0$ and $n^{1/n} \to 1$ $\square$

Convergence of $a^{1/n}$ For $a > 0$, the sequence $a^{1/n}$ converges to 1.

Proof Take $a \geq 1$, and then there is an n such that

$$n \geq a \geq 1.$$

By monotonicity of the nth root function we have

$$n^{1/n} \geq a^{1/n} \geq 1^{1/n} = 1,$$

and by the squeeze lemma with the previous result we have $a^{1/n} \to 1$. Similarly, for $0 < b = 1/a < 1$ we have an n such that

$$\frac{1}{n} \leq \frac{1}{a} \leq 1,$$

$$1 \leq \frac{n}{a} \leq n,$$

$$1 \leq \left(\frac{n}{a}\right)^{1/n} \leq n^{1/n},$$

Now take the limit and with the squeeze lemma we get

$$1 \leq \lim_{n \to \infty} \left(\frac{n}{a}\right)^{1/n} \leq 1.$$

Since $n^{1/n} \to 1$ we must have

$$\lim_{n \to \infty} \left(\frac{1}{a}\right)^{1/n} = 1,$$

and $a^{1/n} \to 1$ for $a > 0$ $\square$

6.4   Power Series

We previously defined power series at the beginning of Section 5.4. For
simplicity and clarity we consider power series about zero, but in all that
follows x could be replaced with $(x-c)$ and not affect the results. Recall
that a power series is

$$\lim_{n\to\infty} \sum_{i=1}^{n} a_i x^i = \sum_{i=1}^{\infty} a_i x^i.$$

The concept of convergence for a power series is the same as for a series.
We require that the sequence of partial sums converge to zero. For

$$S_n(x) = \sum_{i=1}^{n} a_i x^i,$$

we must have for some $b \in \mathbb{R}$

$$\lim_{n\to\infty} S_n(x) = b.$$

Intuitively, the terms in the series must become very small, and faster
than they do for the harmonic series, $\sum 1/n$ which diverges. This could
happen in three ways

- A restriction on the values of x and on the values of a_n.

- A restriction only on the values of a_n.

- Require that $x = 0$.

Previously, in the section on decimal expansions, we derived the geometric sum formula, which is what we have when all the a_i have the same
value a, and $x = r < 1$. So conceptually we start with that as a possible
upper bound on the terms.

$$\sum_{n=0}^{\infty} a_n x^n \leq \lim_{n\to\infty} \frac{a - ar^n}{1-r} = \frac{a}{1-r}.$$

But it isn't quite right because the restriction $x < 1$ is arbitrary and we
would like to find the rule that yields convergence for the most values of

x. Observing that there is a power of n on the left, and that each term added in the series must be less than one for there to be convergence, we bound the series with the geometric series taking c^n as the bound with $c < 1$. This choice allows for an interaction between the x-values and the constants, a_n. Thus we have

$$a_n x^n \le c^n.$$

On the other hand, any finite number of terms, $a_n x^n$ could be greater, and we would still have convergence. So we only require this for sufficiently large $n > N$. Now we'll bound from below as well, and do the algebra

$$-c^n \le a_n x^n \le c^n,$$
$$(|a_n|)^{1/n} |x| \le c,$$
$$|x| \le \frac{c}{(|a_n|)^{1/n}}.$$

This has worked out nicely because $\lim_{n \to \infty} |a|^{1/n} = 1$ for any fixed $a > 0$. Incorporating the fact that we chose $c < 1$ yields

$$|x| \le \frac{c}{(|a_n|)^{1/n}} < \frac{1}{|a_n|^{1/n}}.$$

The problem remains that $|a_n|$ might not converge as a sequence, yet remains bounded. In this case there is a convergent subsequence. To find the largest convergent subsequence we use the limsup. Thus the power series will converge for the values of x that satisfy

$$|x| \le \frac{1}{\limsup_{n > N} |a_n|^{1/n}} = R.$$

Note that within the radius of convergence R, N does not depend on x, and so the convergence is uniform. When $R = 0$ the series only converges at $x = 0$, and when $R = \infty$, the series converges for all values of x. Also note that the above work demonstrated absolute convergence for power series. Since the convergence is uniform, and a series is a sequence of functions (the partial sums, which are each integrable), we have that the series may be integrated term by term. For differentiation we need to apply the interchange of limit and derivative theorem in order to conclude that we can differentiate power series term by term in order to get the derivative. We need to establish the uniform convergence of the derivatives, and we do that next.

Derivatives of partial sums of power series converge uniformly   Suppose that

$$\sum_{n=0}^{\infty} a_n x^n$$

converges for $x \in (R_1, R_2)$. Then

$$\sum_{n=0}^{\infty} n a_n x^{n-1}$$

converges for $x \in (R_1, R_2)$.

Proof   By the given convergence we have that

$$|x| \le \frac{1}{\limsup_{n>N} |a_n|^{1/n}} \le \min\{R_1, R_2\}$$

which is uniform convergence. Since $n^{1/n} \to 1$ we have that

$$|x| \le \frac{1}{(\lim_{n\to\infty} n^{1/n})(\limsup_{n>N} |a_n|^{1/n})} = \frac{1}{\limsup_{n>N} |n a_n|^{1/n}} = \min\{R_1, R_2\}.$$

Therefore

$$\sum_{n=0}^{\infty} n a_n x^{n-1}$$

converges uniformly for $x \in (R_1, R_2)$ □

Differentiate power series term by term   Suppose

$$f_k(x) = \sum_{n=0}^{k} a_n x^n.$$

Then within the radius of convergence $f_k \to f$ uniformly. Also, $f'_k \to g$ uniformly (by the theorem above). Therefore, by the interchange of limit and derivative theorem, $f'_k \to f'$ □

Now that we have rigorously established all the techniques we used in the sine function example from section 5.4, we'll do another example of the same technique to define the exponential function.

Exponential function Solve the following differential equation:

$$f'(x) = f(x) \text{ and } f(0) = 1.$$

Our guess is

$$f(x) = \sum_{n=0}^{\infty} a_n x^n$$

and since we have rigorously justified differentiating term by term

$$\sum_{n=0}^{\infty} n a_n x^{n-1} - \sum_{n=0}^{\infty} a_n x^n = 0,$$

$$\sum_{n=0}^{\infty} (n+1) a_{n+1} x^n - \sum_{n=0}^{\infty} a_n x^n = 0,$$

$$\sum_{n=0}^{\infty} ((n+1) a_{n+1} - a_n) x^n = 0,$$

$$a_{n+1} = \frac{a_n}{n+1}.$$

Applying the initial condition yields

$$f(0) = \sum_{n=0}^{\infty} a_n 0^n = a_0 = 1$$

so that $a_0 = 1, a_1 = 1/2, a_2 = 1/6, \ldots a_n = 1/n!$ Therefore

$$f(x) = \sum_{n=0}^{\infty} \frac{x^n}{n!}$$

which is uniformly convergent, differentiable and integrable term by term, and has derivatives of all orders. We'll use the common notation, e^x without specifying the value of e which you now have the formula to compute.

Those are some very nice properties for a solution to a differential equation, when the technique works. It doesn't always work. Next we'll demonstrate a basic property of the exponential function using this definition.

Exponential function property   Prove $e^{x+y} = e^x e^y$.

Proof   Starting with e^{x+y} we have

$$e^{x+y} = \sum_{n=0}^{\infty} \frac{1}{n!}(x+y)^n.$$

Recall the binomial theorem which states

$$(x+y)^n = \sum_{k=0}^{n} \frac{n!}{k!(n-k)!} x^{n-k} y^k,$$

and apply it to get

$$e^{x+y} = \sum_{n=0}^{\infty} \frac{1}{n!}(x+y)^n = \sum_{n=0}^{\infty} \frac{1}{n!} \sum_{k=0}^{n} \frac{n!}{k!(n-k)!} x^{n-k} y^k,$$

$$e^{x+y} = \sum_{n=0}^{\infty} \sum_{k=0}^{n} \frac{x^{n-k}}{(n-k)!} \frac{y^k}{k!}.$$

Note that k goes from zero to ∞ so $y^k/k!$ passes through the inside sum:

$$e^{x+y} = \sum_{n=0}^{\infty} \frac{y^n}{n!} \sum_{k=0}^{n} \frac{x^{n-k}}{(n-k)!}.$$

Also, when $n = 0$, $k = n = 0$ so the sum on the right is only $\sum_{n=0}^{\infty} x^n/n!$ and we have

$$e^{x+y} = \sum_{n=0}^{\infty} \frac{y^n}{n!} \sum_{n=0}^{\infty} \frac{x^n}{n!} = e^y e^x \quad \square$$

It should be short work to show that the sum of two convergent power series is yet another power series, convergent on their common radius of convergence. Next we'll present the proof for products of power series.

Product of two power series is a power series   Let $f(x) = \sum_{n=0}^{\infty} a_n x^n$ and $g(x) = \sum_{m=0}^{\infty} b_m x^m$ be convergent power series on the common interval (c, d). Then

$$f(x)g(x) = \sum_{k=0}^{\infty} c_k x^k$$

is convergent on (c, d).

Proof Outline   We start by writing the product of the power series:

$$\sum_{n=0}^{\infty} a_n x^n \sum_{m=0}^{\infty} b_m x^m.$$

Since the convergence of power series is absolute, we can interchange limits on the common interval of convergence as proven in the section on series:

$$= \sum_{m+n=0}^{\infty} a_n b_m x^{m+n}.$$

To demonstrate that this is indeed in the form of a power series we write out the coefficients

$$a_0 b_0 + a_0 b_1 + a_0 b_2 + a_0 b_3 + \cdots + a_0 b_k +$$

$$a_1 b_0 + a_1 b_1 + a_1 b_2 + a_1 b_3 + \cdots + a_1 b_k +$$

$$a_2 b_0 + a_2 b_1 + a_2 b_2 + a_2 b_3 + \cdots + a_2 b_k +$$

$$\vdots$$

$$a_k b_0 + a_k b_1 + a_k b_2 + a_k b_3 + \cdots + a_k b_k$$

which can be much more concisely written as

$$\sum_{k=0}^{\infty} c_k = \sum_{k=0}^{\infty} a_0 b_k + a_1 b_{k-1} + \cdots + a_k b_0.$$

Then with $k = m + n$ we have

$$\sum_{n=0}^{\infty} a_n x^n \sum_{m=0}^{\infty} b_m x^m = \sum_{k=0}^{\infty} c_k x^k \quad \square$$

We don't prove that f/g is a power series for $g \neq 0$, although it will be, but possibly under a smaller interval. In order to understand the phenomena we introduce the tool of Taylor series first. Then an example follows to illustrate why our study of series needs to move to another domain (literally).

6.5 Taylor Series

Often the Taylor series is introduced as a way to approximate a function when we already have a rule to compute the function. This begs the question, "What's the point?" The usual answer is probably something about how a calculator computes the value that it reports. Another perspective comes from driving a car. We have an instrument in the car, called a speedometer, that tells us the instantaneous velocity, or first derivative of position. Smartphones have an accelerometer in them that tells us the instantaneous acceleration as well. What happens if we develop an approximation of a function by using the derivatives at a particular point? We get the Taylor series. This is useful because there is often instrumentation (more elaborate than a speedometer and a smartphone) that collects data about a process we want to model. Rather than just give the definition, we'll build up the idea. Inspired by the mean value theorem we know that for any function with a derivative we have a line with the same derivative at some point x_0

$$T_1(x, x_0) = f(x_0) + f'(x_0)(x - x_0).$$

Then if in addition we want our next approximation $T_2(x, x_0)$ to have $T_2''(x, x_0) = f''(x_0)$ and be a polynomial we must have something like

$$T_2(x, x_0) = f(x_0) + f'(x_0)(x - x_0) + \frac{f''(x_0)(x - x_0)^2}{2}.$$

If we continue this process of requiring each successive polynomial approximation to have the same derivative as the original function we inductively arrive at

$$T_n(x, x_0) = f(x_0) + f^{(1)}(x_0)(x - x_0) + \frac{f^{(2)}(x_0)(x - x_0)^2}{2} + \cdots + \frac{f^{(n)}(x_0)(x - x_0)^n}{n!},$$

where $f^{(n)}$ is the n-th derivative and we are assuming that f has n derivatives. This is more compactly written in the usual form as

$$T_N(x, x_0) = \sum_{n=0}^{N} \frac{f^{(n)}(x_0)}{n!}(x - x_0)^n,$$

where $f^{(0)} = f$. The the question remains, how good is the approximation? We answer this question by studying

$$R_n(x) = f(x) - T_n(x, x_0),$$

when we are trying to determine where $f(x)$ will be next. Think about the car scenario above where x would be time. If we have the first and second derivative at a particular point in time, we ought to be able to determine approximately where we will be in a second or two. That is, we want to use $T_n(x, x_0)$ but at a point t different from x_0. So we consider

$$g(t) = T_N(x, t) = \sum_{n=0}^{N} \frac{f^{(n)}(t)}{n!}(x - t)^n.$$

Now

$$R_n(x) = f(x) - g(t),$$

but it happens, due to all the higher order terms canceling out in the Taylor series, that we have $f(x) = g(x)$. To see this, apply the fundamental theorem of calculus

$$R_n(x) = g(x) - g(t) = \int_t^x g'(y) dy.$$

Then if we apply the definition of g and differentiate before we integrate we find a telescoping series as follows:

$$= \int_t^x f'(y) - f'(y) \quad + f''(y)(x - y)$$

$$-f''(y)(x - y) \quad + \frac{f'''(y)(x-y)^2}{2}$$

$$-\frac{f'''(y)(x-y)^2}{2} + \frac{f^{(4)}(y)(x-y)^3}{3!} \cdots$$

Despite the $\cdots$ at the end, it is a finite sum $T_N(x, t)_t$ and we have

$$R_n(x) = \int_t^x \frac{f^{(n+1)}(y)(x - y)^n}{n!} dy.$$

An easy to apply bound is to choose an M so that $|f^{(n+1)}(y)| \leq M$ and we then have

$$|R_n(x)| \leq \frac{M(x - t)^{n+1}}{(n + 1)!}.$$

Now suppose our car is traveling at 7 meters per second at time zero with acceleration of 5 meters per second squared, and a jerk (the third derivative of position) of less that 2 meters per second cubed. What will be our approximate position along the straight line we are traveling over the next second?

$$T_2(t,0) = 0 + 7t + \frac{5t^2}{2},$$

$$|R_n(x)| \leq \frac{M(x-t)^{n+1}}{(n+1)!} = \frac{2t^3}{(3)!}.$$

So that the position is given by

$$f(t) = 7t + \frac{5t^2}{2} \pm \frac{t^3}{3} \quad \text{meters from the origin,}$$

and plotted below, with the error bounds dotted.

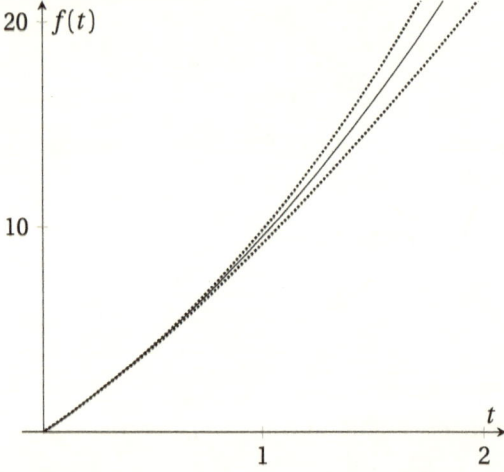

Notice that in this Taylor series problem the true function remains unknown, yet we are able to draw conclusions about the function with other information. The essential difference between a power series and a Taylor series has to do with the number of derivatives that might be available, and how this affects the radius of convergence. In the discussion on power series, we showed that differentiating a power series yields another power series with the same radius of convergence. The way to write this concisely is $f \in C^\infty$, f is in the set of functions with continuous derivatives of all orders. In order to compute the nth Taylor polynomial we only required that $f \in C^n$.

A natural question at this point would be about the relationship between power series and Taylor series. Wouldn't it be nice if they turned out to be the same series on the same interval for a given function? Yes, it is nice when that happens. Functions with this property are called analytic. But for this to be the case we must have $R_n(x) \to 0$ which doesn't always happen. Next we'll give a few definitions, and then an example to illustrate the definitions.

Real Analytic Function A real function is analytic at a point $x_0 \in (a, b)$ if $f \in C^\infty$ with $R_n(x) \to 0$ for $x \in (a, b)$.

If we think about the implications of this in terms of the point where the function is analytic, the definition tells us that the remainder converges to zero for any other point in the interval of convergence as well. So a function that is analytic at one point in the interval of convergence (for the Taylor series) is analytic at all points in the interval of convergence. This is more generally called analytic continuation, and motivates the next definition of analytic.

Analytic function A function with a power series expansion about all points in the function's domain.

Multiplicative and additive closure of analytic functions Suppose f and g are analytic. Then $f \pm g$ and $f \cdot g$ are analytic.

Since analytic functions are (for practical purposes) convergent power series, the closure was almost established at the end of the last section. So what's the rub? The next example illustrates the problem with division of convergent power series, which can cause a drastic reduction in the interval of convergence that has no reasonable explanation in the world of real variables. For the remainder of the book, much of the computational detail is omitted so we can focus on the ideas, and results. The reader should have little trouble filling in the gaps by this point.

Example 49 Let $f(x) = 1$ and $g(x) = x^2 + 1$. Then f/g has a severely reduced radius of convergence even though $g(x) \neq 0$. Note that f is a power series with $a_1 = a_2 = \cdots = 0$ and similarly for g. A few pages of scartch work will get you the Taylor series for f/g:

$$T_N(x,0) = 1 - x^2 + x^4 - x^6 + x^8 \cdots = \sum_{n=0}^{N} (-1)^n x^{2n}$$

An application of the ratio test yields that

$$|x^2| < 1$$

is required for convergence, so that the radius of convergence is one. Below is a picture of what is going on:

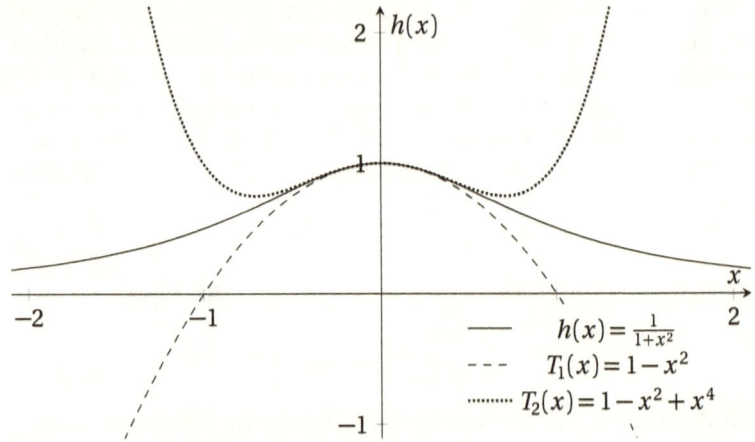

$h(x)$

-2 -1 2

$\qquad\qquad h(x) = \frac{1}{1+x^2}$
$\text{- - -}\quad T_1(x) = 1 - x^2$
$\cdots\cdots\; T_2(x) = 1 - x^2 + x^4$

We have nice convergence in the radius, but outside the radius the series diverges wildly for any value of x, as can be seen in the computation of the remainder:

$$R_n(x) = \frac{1}{n!} \int_0^x (2n)!(t-0)^n \, dt = \frac{(2n)!}{(n+1)!} x^{n+1}$$

This seems to be unreasonable behavior for a series representation of a rational function with non-zero denominator. But that is where the problem is. In order to account for what is happening we observe that $\frac{1}{1+x^2}$ is dividing by zero when $x = \sqrt{-1}$. For this reason the proper definition of an analytic function can only really be given in terms of complex variable theory.

Analytic function A function $f(z): \mathbb{C} \rightarrow \mathbb{C}$ is analytic at z_0 if it is differentiable at $z_0 \in \mathbb{C}$.

The requirement for differentiability in the complex plane is much more restrictive because z may approach z_0 from any direction. On the other hand, when we compose a real analytic function with one that isn't analytic, we naturally don't necessarily get an analytic function. There is a large class of functions made this way that is very useful for modeling point sources (of force, electrical charge, mass, etc.). The next example shows that the Taylor series of the composed function converges, but not to the original function.

Example 50 Let $f(x) = e^x$ and $g(x) = 1/(x^2-1)$. Then $h(x) = f \circ g$ is not analytic at $x = 0$ because there is not an open interval containing 0 with $R_n(x) \rightarrow 0$. Below are the first few derivatives that go in the Taylor series.

$$h'(x) = e^{\frac{1}{x^2-1}} \frac{-2x}{(x^2-1)^2}$$

$$h''(x) = e^{\frac{1}{x^2-1}} \frac{6x^4-2}{(x^2-1)^4}$$

$$h'''(x) = e^{\frac{1}{x^2-1}} \frac{-4x(6x^6+3x^4-10x^2+3)}{(x^2-1)^6}$$

$$h^{(4)}(x) = e^{\frac{1}{x^2-1}} \frac{4(30x^{10}+45x^8-132x^6+58x^4+6x^2-3)}{(x^2-1)^8}$$

Notice that each of the derivatives is finite at $x = 0$. Then

$$T_N(x,0) = \sum_{n=0}^{N} \frac{h^{(n)}(0)}{n!}(x)^n,$$

and at $x = 0$

$$\lim_{N \to \infty} \sum_{n=0}^{N} \frac{h^{(n)}(0)}{n!}(0)^n = 0.$$

Therefore

$$\lim_{N\to\infty} R_N(0) = \lim_{N\to\infty} h(0) - T_N(0,0) = h(0) = \frac{1}{e},$$

and $h(x)$ is not analytic at $x = 0$. Below is a plot of $h(x)$ on (-1,1):

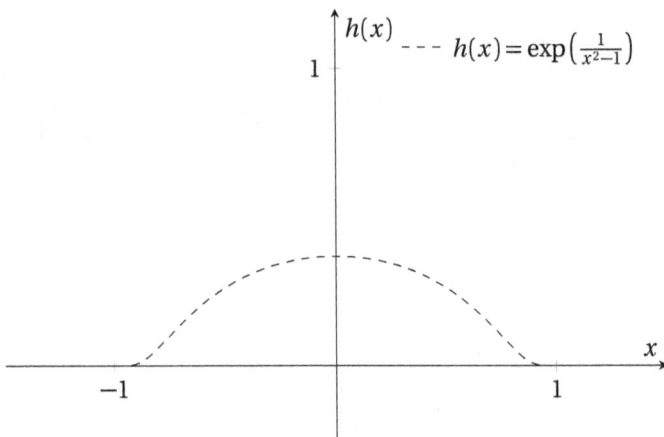

Notice that $\lim_{x\to 1} h(x) = 0$ so that the extension

$$\bar{h}(x) = \begin{cases} \exp\left(\frac{1}{x^2-1}\right) & x \in (-1,1) \\ 0 & x \notin (-1,1) \end{cases}$$

is continuous. This is an example of a function with compact support, a function whose value is zero outside some compact set. We can change the shape of the function in various ways. In the plots below we define the function as zero outside the root of the denominator in the exponent.

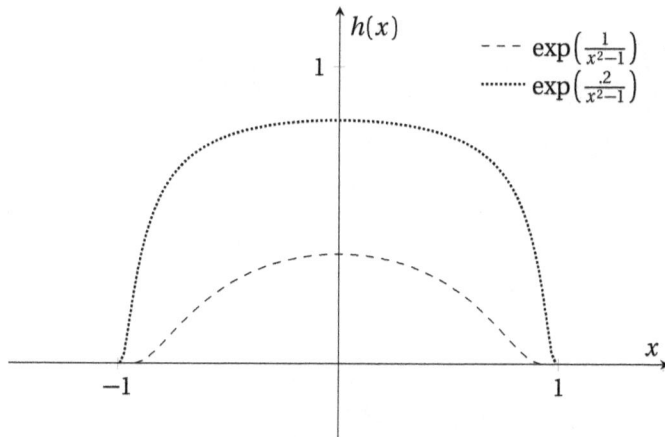

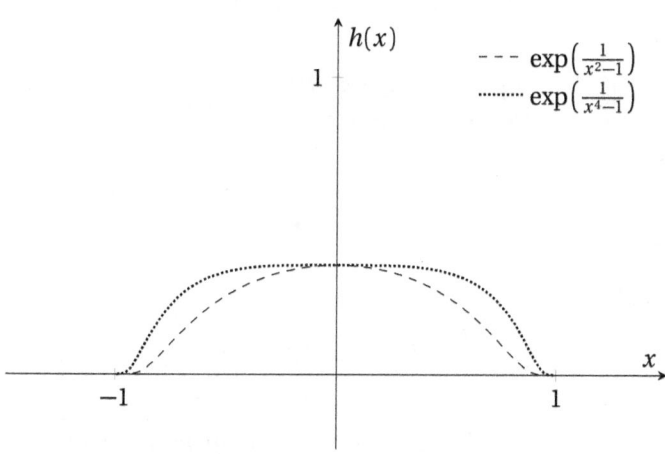

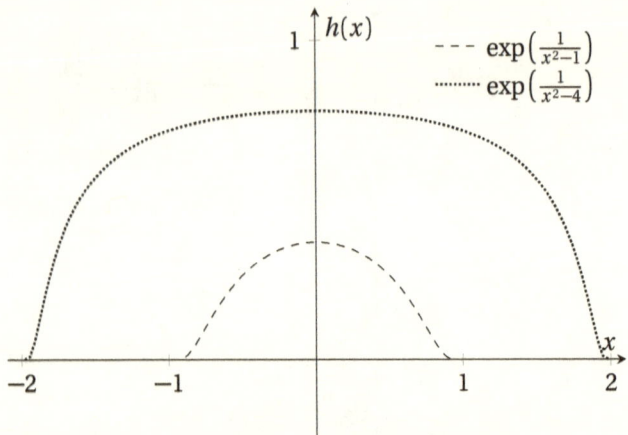

These are all examples of test functions for distributions, or for general-ized functions. They might also be called blip functions in some litera-ture. They provide a rigorous justification for the use of the delta "func-tion," $\delta(x)$. The quotes are there because it isn't really a function, but it has this nice property for any other function, f:

$$\int_{-\infty}^{\infty} \delta(x)f(x)dx = f(0).$$

And so shifting the delta function by a as in $\delta(x-a)$ gives considerable flexibility:

$$\int_{-\infty}^{\infty} \delta(x-a)f(x)dx = f(a).$$

When $f(x) = 1$ we have

$$\int_{-\infty}^{\infty} \delta(x)dx = 1,$$

but in practice the convention is that

$$\delta(x) = \begin{cases} \infty \text{ (or undefined)} & x = 0 \\ 0 & x \neq 0 \end{cases}$$

which is not consistent with our definitions of function and integration. As a result, Laurent Schwartz developed the theory of distributions. One way to think about them is to consider a sequence of test functions, $\phi_n(x)$ (like the ones described above) such that for each n we require

$$\int_{\frac{-1}{n}}^{\frac{1}{n}} \phi_n(x)dx = 1,$$

and $\phi_n(x) = 0$ outside $(-1/n, 1/n)$. Then define

$$\delta(x) = \lim_{n \to \infty} \int_{-\infty}^{\infty} \phi_n(x)dx = 1.$$

The details for the construction, and applications that inspired the use of generalized functions could easily fill an entire college course.

Conclusion   Here in this final chapter we have discovered two directions for further study. One being analytic functions, which is in a course usually called complex variables, or complex analysis. The other, generalized functions or distributions, would be part of an advanced course in partial differential equations. Other directions would be the course in real analysis to study measure theory and Lebesgue integration, or an introductory course in topology to fill in the gaps that we glossed over in chapter three.

Hopefully this exposition has been a useful supplement to a first course in mathematical analysis by highlighting conceptual links, and presenting some more intuition about analysis proofs concisely, but in a way useful to the student.

Bibliography

[1] Colin Adams and Robert Franzosa, *Introduction to Topology: Pure and Applied*, Pearson Education, New Jersey, 1st edition, 2008.

[2] Mark J Albowitz & Athanassios S Fokas, *Complex Variables: Introduction and Applications*, Cambridge University Press, United Kingdom, 1st edition, 1997.

[3] Robert G. Bartle, *The Elements of Real Analysis*, John Wiley and Sons, New York, 2nd edition, 1976.

[4] David M. Bressoud, *A Radical Approach to Lebesgue's Theory of Integration*, Cambridge University Press, New York, 1st edition, 2008.

[5] R. Creighton Buck, *Advanced Calculus*, Waveland Press, Illinois, 3rd edition, 1978.

[6] Patrick M. Fitzpatrick, *Advanced Calculus: A Course in Mathematical Analysis*, PWS Publishing Company, Massachusetts, 1st edition, 1996.

[7] John B. Fraleigh, *A First Course in Abstract Algebra*, Addison Wesley, Massachusetts, 7th edition, 2003.

[8] Bernard R. Gelbaum and John M.H. Olmsted, *Counterexamples in Analysis*, Dover, New York, reprint of 1964 edition, 1992.

[9] Michael T. Heath, *Scientific Computing: An Introductory Survey*, McGraw Hill, New York, 2nd edition, 2002.

[10] John H. Holland, *Hidden Order: How Adaptation Builds Complexity*, Basic Books, New York, 1st edition, 1995.

[11] A.N. Kolmogorov & S.V. Fomin (Translated by Richard A. Silverman), *Introductory Real Analysis*, Dover, New York, edited translation of 1968 Moscow edition, 1970.

[12] James R. Munkres, *Topology*, Prentice Hall, New Jersey, 2nd edition, 2000.

[13] Kenneth A. Ross, *Elementary Analysis: The Theory of Calculus*, Springer, New York, 2nd edition, 2013.

[14] Gilbert Strang, *Calculus*, Wellesley-Cambridge Press, Massachusetts, 2nd edition, 2010.

[15] Robert S. Strichartz, *The Way of Analysis*, Jones and Bartlett, Massachusetts, revised edition, 2000.

Index